FOR THE LOVE OF M

# FOR THE LOVE OF MARS

## A HUMAN HISTORY OF THE RED PLANET

### MATTHEW SHINDELL

THE UNIVERSITY OF CHICAGO PRESS

CHICAGO AND LONDON

The University of Chicago Press, Chicago 60637
The University of Chicago Press, Ltd., London
© 2023 by Smithsonian Institution

Published 2023
Printed in the United States of America

32  31  30  29  28  27  26  25  24  23      1  2  3  4  5

ISBN-13: 978-0-226-82189-4 (cloth)
ISBN-13: 978-0-226-82190-0 (e-book)
DOI: https://doi.org/10.7208/chicago/9780226821900.001.0001

Library of Congress Cataloging-in-Publication Data

Names: Shindell, Matthew, author.
Title: For the love of Mars : a human history of the red planet / Matthew Shindell.
Description: Chicago ; London : The University of Chicago Press, 2023. | Includes bibliographical references and index.
Identifiers: LCCN 2022035682 | ISBN 9780226821894 (cloth) | ISBN 9780226821900 (ebook)
Subjects: LCSH: Astronomy—History. | Mars (Planet) | Mars (Planet)—Exploration—History.
Classification: LCC QB641 .S487 2023 | DDC 523.43—dc23/eng20221021
LC record available at https://lccn.loc.gov/2022035682

♾ This paper meets the requirements of ANSI/NISO Z39.48-1992 (Permanence of Paper).

FOR JEANNETTE

# CONTENTS

# *ILLUSTRATIONS*

**PLATES FOLLOW PAGE 136**

# MARS IN THE TIME OF COVID-19

On July 30, 2020, I tuned my smart TV to NASA's YouTube channel to watch the Mars 2020 Perseverance rover set out on its 314-million-mile journey. I watched the rover, weighing just over one ton, launch into low Earth orbit atop a fiery Atlas V rocket from Cape Canaveral, Florida. I continued to watch via feeds from the onboard cameras as the rover left Earth's gravity behind and set itself in a solar orbit toward a Martian rendezvous. Under normal circumstances I might have tried to see the launch in person. In 2020, however, I didn't even get the chance to try.

I wasn't alone. Because of the coronavirus disease 2019 (COVID-19) global pandemic, most of the engineers and scientists who had designed and built the rover, who had devoted years of their lives and careers to it, also had to stay home. Launch Complex 41 should have been filled with the nervous energy of hundreds of emotionally invested spectators. Instead, the various rover teams met together over Zoom and other remote meeting platforms to watch the launch "together" via a version of the same live digital feed I was watching. Meanwhile, at mission control, launch team members wore special "Mars 2020" facemasks and kept six feet apart from one another as they went through their checklists and counted down to liftoff. NASA communications officers sat separated by clear plexiglass dividers as they provided live commentary and interviews with NASA officials.

During the hour before the launch, commentators explained that similar measures had been taken during the months of preparation ahead of delivering the rover from the cleanroom facility at the Jet Propulsion Laboratory (JPL), where it had been assembled. The assembly team had worked throughout the pandemic even as most of the world, including many of their colleagues at JPL and other NASA centers, observed orders to stay at home. Signs reading "JPL Safe at Work" could be seen posted on walls around the JPL campus, reminding employees that these were not normal times. Perseverance—a name chosen by a thirteen-year-old student from Virginia—came to mean something very specific to the team as they worked to meet an unforgiving deadline set by the regular orbits of Earth and Mars.[1]

The stress level at JPL and the Kennedy Space Center (KSC) was no doubt extra high, coming not only from the pressures of meeting testing and assembly deadlines, but from the daily reports of new infections and deaths from the virus. "It really began to affect us in mid-March [2020]," Perseverance deputy project manager Matt Wallace said during a news conference. "We were at a critical time in the processing for the spacecraft. All of the elements were down at Kennedy Space Center, and we had to fully assemble and do the final testing of the spacecraft. It had to be done right—you can't make a mistake at that point—and of course the environment made that a lot more difficult."[2]

It is not uncommon for spacecraft teams to add something of special significance to the spacecraft as it is being prepared for delivery. In 2003, JPL engineers affixed a plaque to the Spirit rover commemorating the crew of the space shuttle Columbia (STS-107), tragically lost in February of that year. This time, the Perseverance rover team's minds turned to the health care workers who had put themselves on the frontlines in the pandemic, risking their lives for the well-being of their fellow citizens. They designed a three-by-five-inch plate that combined the familiar image of the medical staff with a likewise familiar image of Earth's Western Hemisphere, including the Perseverance rover's launch from Florida and orbital departure toward Mars (figure 1).

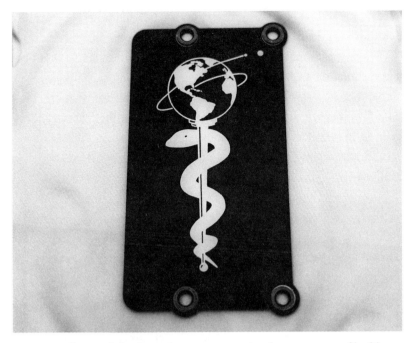

**FIGURE 1** Illustrated aluminum plate commemorating the perseverance of health care workers around the world during the COVID-19 pandemic. This plate was affixed to the Perseverance rover and sent to Mars. NASA/JPL-Caltech.

Watching this all come together, it occurred to me that some of the ideas illustrated on this plaque would be familiar to the ancient Greeks or to medieval European natural philosophers. The serpent-entwined rod regularly used today to represent the medical community—a version of which dominates the plaque—is, after all, based on the symbology of Asclepius, an ancient Greek god associated with the medical arts. Along with its double-serpented counterpart, the caduceus (itself a symbol of the Greek god Hermes, and only recently associated with medicine), it has endured for millennia. Our medieval counterparts might not, however, have understood Earth and its relationship to Mars as depicted in this illustration. The idea that a human-made object, forged from metals and other earthly substances, could somehow leave Earth and travel beyond the Moon—or that there would be anything beyond the Moon to land on—would defy their physical understanding of the cosmos. And though they would have recognized a spherical Earth, they certainly wouldn't have recognized the strange continents of North and South America represented prominently in this symbol.

There is a lot medieval Europeans would not understand about the idea of sending a spacecraft to a faraway planet. They would, however, understand the association between Mars and health and illness. They also might agree that a tribute to Mars should be dedicated to those willing to sacrifice themselves for a greater good. And they definitely would have agreed that during a pandemic, it is important to look to the heavens.

### MY JOURNEY TO MARS

Before the pandemic began, I already knew I was writing a book about Mars. But my original idea for this book was to focus on Mars in the twentieth and twenty-first centuries—the time of robotic Mars exploration. One chapter would look at pre–space age ideas about Mars, but the bulk of the book would address how we've explored Mars from the Cold War era until today. Individual chapters would be devoted to the spacecraft that had flown by, orbited, landed, and roved on the planet. But something changed.

On March 14, 2020, the Smithsonian sent my colleagues and me home. Like so many other people who had worked in offices, we worked from home for the next several months thinking that our return to the office was imminent. It wasn't. It became obvious that this extended period of telework would last for years, not months. I couldn't travel or visit archives. I couldn't even go out to dinner. But I could read.

I decided to take a deep dive into the history I had planned to use just as prelude to what I thought were the main events in the history of Mars. Other Mars books spent one or two sentences summing up ancient beliefs about Mars, but what did these beliefs mean in context? Who actually cared about Mars in ancient cultures, and why? Had people really feared Mars and treated it as a bad omen?

And what about the Middle Ages? I understood this period's importance in laying the groundwork for the emergence of "modern science," but my knowledge of the medieval cosmos was superficial. I knew more about the Renaissance, but my perspective was still limited. As I began each foray into a period's ideas about Mars, I found myself out of my depth. I tried to understand Mars and its relationship to the world in which my subjects had lived, from their perspective. And so I kept reading, chasing footnotes from one book to another, until I felt I could write a chapter that told the story of Mars during that period responsibly.

The first finished draft of this book was full of tangents. I was excited to include every piece of information that excited me when I first encountered it. These chapters took "the scenic route" to Mars. Fortunately, enough time passed between the completion of the first full draft and my final manuscript that I was able to let go of a lot of these intellectual side quests. What I am left with, I hope, is a streamlined but nuanced book about the history of human ideas about Mars. It is by no means complete—there are cultures I still know too little about to approach. Nor is it comprehensive—I would love to spend more time with the original texts from these periods, so that I can better inhabit their worlds. I hope that time will allow me to add some of these stories to a later project, or to an expanded version of this one.

This approach to Mars did reshape the questions I ultimately wanted to pose in this book. I originally imagined that the overarching questions would be "why?" questions. Why do we want to explore Mars and even send humans? Why did we see Mars one way during one century, only to shift our understanding dramatically the next? But in the end, developing this book as a "long" history of Mars transformed these into "who?" questions: Who cared about Mars in the past? Who cares about it today? Who are we that we should go to Mars? The "who" question—how we understand ourselves, our place in the universe, and what we hope to become—will shape why we go to Mars and what we do there. How we understand our history is equally important. Should we talk in terms of "colonizing" Mars, knowing the history of colonialism on Earth? Does this language limit our ability to imagine a future for Mars that includes everyone?

Of course the "we" of Mars exploration will be construed differently by different readers, and it is my hope that by the conclusion of this book it will be obvious that the "we" should be construed as broadly as possible. Our discussion of who we are should include everyone. Our discussion of our history should likewise not center those perspectives that have historically been privileged (though they do still tend to dominate many of the forums where spaceflight is discussed and planned). We might not all agree on who we are and what we want for our future— or for Mars's future—but we can all participate in asking these questions together.

# KEEPING UP WITH MARS

In the fall of 2020, three nations launched robotic explorers to Mars. The United States sent its fifth rover, Perseverance, the latest in an impressive line of successful spacecraft missions. China's Tianwen 1 (Heavenly Questions) mission was launched as the Chinese space program's second attempt to reach the planet, this time with an ambitious trio of orbiter, lander, and rover. And the United Arab Emirates' Al-Amal (Hope) orbiter flew as the UAE's first attempt. The history of Mars exploration—one in which missions are as likely to fail as not—says that Mars eats spacecraft. Remarkably, all of these missions arrived at Mars operational.

The 2020 launch window wasn't a complete triumph. A fourth mission from the European Union and Russia didn't make it to the launchpad. The life-seeking rover, named in honor of the British crystallographer Rosalind Franklin, had to be postponed; faced with an already tight schedule exacerbated by a global pandemic, the European Space Agency and Roscosmos decided their joint mission could wait. When Russia invaded Ukraine in the spring of 2022, these plans were canceled altogether, leaving the EU's rover without a ride to Mars.

Because Mars takes about twice as long as Earth to complete an orbit around the Sun, opportunities to send spacecraft to the red planet happen only roughly once every two years, when Earth overtakes it on the inside track and the two planets are near each other in their orbits, lined up with the Sun in an arrangement astronomers call "opposition." Not all oppositions are equally favorable—planetary orbits are elliptical, so the distance between Mars and Earth in opposition can differ dramatically. An opposition during perihelion (when Mars is closest to the Sun in its orbit) can nearly halve the distance between the two planets from sixty- to thirty-odd million miles (the precise numbers vary from century to century). These closer oppositions occur about once every sixteen years, and the 2020 launch window was one of these encounters, with Mars only 38.57 million miles distant.

A close opposition makes it easier to send something to Mars, but it doesn't make it cheap. Mars exploration always costs a good amount of money; development of Perseverance cost roughly $2.2 billion, along with an additional nearly quarter billion dollars for launch. Operating costs over the next five to ten years of the mission will cost millions more. One might ask—and many do—what is so valuable about Mars exploration?

Some will say we explore Mars because it's in our nature: we have an innate drive to explore new places, and this has been an instrumental force in shaping our history on this planet. But I don't find this answer very convincing. This narrative of Mars as the next frontier, essential to our progress as a species, tends to be uncritical of the actual history of European exploration and colonial expansion, or dismissive of its destructive and genocidal tendencies. It is also historically inaccurate. As this book will show, for most of human history our interest in Mars had little to do with exploration. It has had much more to do with how we perceived ourselves, our world, and our connection to Mars.

Even today, when we can legitimately say we are exploring the surface of Mars, we do not do this simply for the sake of exploration. The exploration of the solar system over the past sixty years has contributed to advances in knowledge about Earth's history, the processes that shaped it, and its likely future. In its study of one of our nearest neighbors in the so-called habitable zone, Mars science has played a key role in advancing knowledge about Earth. As we have learned more about Mars and the geological path it has taken, we have gained important insights into what makes Earth an oasis in an otherwise hostile and possibly desolate solar system, as well as what changes might put life as we know it here at risk.

But, of course, it's not just about the science.

Chances are, if you watch television or movies, that you've already noticed ways in which Mars exploration gets connected to multiple narratives concerning the human future. What we learn about planetary processes through our study of Mars may help us better understand and even combat problems like climate change. I believe that any responsible approach to Mars exploration should prioritize the development of knowledge and technologies that will help mitigate climate change, even if indirectly.

But in other increasingly popular visions of the role Mars might play in saving us from ourselves, Mars and the technologies we might use to make parts or the entire planet habitable for humans become an escape hatch through which we—or some privileged few—can be saved. One such proposal advocates that Mars should be terraformed (transformed on a planetary scale) into a second Earth, where humans will be protected from any human-made catastrophe that befalls Earth. In my view, this is like raising the Titanic because you need a boat. Maybe it can be done, but is it the best solution? If we want to save the humans of Earth, shouldn't we focus on Earth? There may be good reasons to terraform Mars, and benefits to becoming a multiplanet species, but it shouldn't stop us from directing our resources to solving climate change. It would be ridiculous to think that making an uninhabitable planet habitable would be easier than preserving life on our own.

Mars exploration is part of another more immediate legacy: that of the mobilization of science and technology during the Cold War. Our first robotic and human expeditions beyond Earth were conducted as part of a Cold War competition between the United States and the Soviet Union. These missions were operated as civilian programs, but the technologies developed to launch spacecraft were essentially the same as those developed for ballistic missiles. And while we call it a "cold" war because it somehow never became a hot or explosive conflict, the truth is that it was far from bloodless and had real consequences for the world we live in today. It shaped postcolonial relationships between the global north and south, cost lives and political stability in South America and Asia, and left the toxic detritus of atomic testing and weapons proliferation. Even today, the technologies we use to study Mars are connected not only to the global communications networks so central to our everyday lives, but also to dark networks of spying, surveillance, and military readiness.

To further unpack the reasons why we go to Mars, let's very briefly examine the motivations of the three missions launched in 2020. For the moment, let's set aside the scientific questions these missions are meant to answer.

### WHY PERSEVERANCE WENT TO MARS

When a rover lands on Mars, it does demonstrate the incredible feats nations can accomplish when government funding, political will, and technological know-how are aligned with national priorities. Among other things, it demonstrates an impressive level of technocratic ability and organizational power. For the United States, sending missions to Mars has typically been an occasion for great fanfare. These occasions are by design meant for global consumption, as are the events surrounding arrival at Mars and the descent to the surface.

During the Cold War, these moments were one way of providing a civilian face for space technology and infrastructure while demonstrating the capabilities of these technologies through peaceful means. Even though the Cold War context for space exploration no longer exists, these displays of technoscientific gymnastics have not become irrelevant. The United States continues to use space as an arena in which to show off its economic and technological might, to build alliances, and to compete with adversaries. These moments are still used to legitimate political claims of global leadership—in space and on Earth.

For NASA, the agency tasked with carrying out these public spectacles, two endeavors have proved to be particularly awe-inspiring to the public at home and the world at large: human spaceflight and Mars rover missions. Both of these activities are highly visible and allow NASA to participate in (and become central to) conversations about American character, technological skill, innovation, and human progress. In addition to the impressive technology of the robotic rover itself, Perseverance carried an untried technology to Mars: a small autonomous rotorcraft named Ingenuity meant to prove the feasibility of airborne exploration (figure 2). The Perseverance mission is tied to a sequence of missions that will, if carried through, return samples from Mars—a goal that both planetary scientists and human exploration advocates have wanted since the 1970s. It also carries spacesuit materials exposed to the Martian elements and an experimental technology demonstration that pulls oxygen out of the planet's thin carbon dioxide atmosphere to prove that humans on Mars could generate fuel for the

FIGURE 2 "Selfie" of the Perseverance rover and Ingenuity rotorcraft on Mars, taken with the rover's WATSON camera. NASA/JPL.

trip back to Earth. These experiments allow NASA to present itself as leading the way toward eventual human exploration of Mars.

### WHY THE UNITED ARAB EMIRATES WENT TO MARS

On February 9, 2021, the United Arab Emirates became the first Arab country to reach Mars when the Hope probe successfully entered into orbit around the red planet. Seven years earlier, Sheikh Khalifa bin Zayed Al Nahyan, president of the UAE and ruler of Abu Dhabi, and Mohammed bin Rashid Al Maktoum, vice president and prime minister of the UAE and ruler of Dubai, announced that the Emirates Mars Mission would be developed by the Mohammed bin Rashid Space Centre (MBRSC), working in conjunction with international partners and funded by the UAE Space Agency.

The Cold War trained us to think of spaceflight—especially ambitious Mars missions—as something a nation does to signal that it has developed a "space economy" and possesses the expertise and technological capabilities to build and operate rockets, missiles, and spy satellites. Viewed through a geopolitical lens, it is a type of highly publicized technological flexing—a gun show. But this was not quite the case for the UAE.

Emirati engineers designed and built the Hope spacecraft in Boulder, Colorado, in collaboration with the Laboratory for Atmospheric and Space Physics (LASP) at the University of Colorado Boulder, with support from Arizona State University and the University of California, Berkeley. Working side by side with LASP engineers, Hope team members learned skills that will serve them in later missions. Once Hope was completed, a Ukrainian transport plane delivered it to Dubai for testing. Finally, Hope flew to Japan for launch, where the UAE arranged with the Japan Aerospace Exploration Agency (JAXA) for launch on a Mitsubishi Heavy Industries H-IIA rocket from Japan's Tanegashima Island launch facility.

The UAE has made clear that while it's looking forward to doing some good Martian science, this is not the primary reason why the young nation spent $200 million on building and launching the spacecraft. "It's not about reaching Mars," said Hope project manager Omran Sharaf. "It's about getting the ball rolling and . . . changing the mind-set"—shifting UAE priorities from oil to science and technology development that will serve the country in a post–fossil fuel economy.[1] The main goal of the mission is to plant the seeds for a science and technology workforce that can build a future for the country. It was a signal to the world, and even more importantly to the people of the UAE, that new things are coming.

## WHY CHINA WENT TO MARS

When China's Zhurong rover rolled out onto the surface of Mars on May 21, 2021, China became only the second nation in the world to successfully operate a rover on the red planet. Compared to the UAE's Mars mission, China's orbiter, lander, and rover have been interpreted as signals of a more traditional type. As US astronaut Pamela Melroy stated in a congressional hearing after the Zhurong landing, "China has made their goals very clear, to take away space superiority from the United States." She warned, "We are right to be concerned."[2]

Also in 2021, China began construction of a new orbiting space station and announced plans to send taikonauts (the Chinese equivalent of cosmonauts or astronauts) to the orbiting construction site and to the Moon. Only months earlier, on December 16, 2020, samples returned from the Moon by the Chang'e-5 lunar probe were recovered in the grasslands of Siziwang Banner in the Ulanqab region of Inner Mongolia. All of these activities combined do seem to speak to a nation that wants to be taken seriously as a space power.

While the United States and China do not currently find themselves in a "space race" like what characterized the first period of space exploration, they nonetheless do find themselves competing for influence as the new landscape of international and commercial spaceflight takes shape, as they are likewise gripped in an economic and political competition on the ground. Both nations would like to be able to attract and lead international partners in establishing norms for low-Earth-orbit and lunar operations.

Mars exploration has become increasingly global in the post–Cold War period. The European Union, Russia, and India have all flown successful missions to Mars in the twenty-first century, and even more nations are involved in spaceflight more generally. But none of these has typically been perceived to be a threat to US space leadership. This highlights the fact that space politics are intimately connected to politics on the ground. Space is a place for peaceful cooperation—except when it's an arena for establishing geopolitical hegemony. The US reaction to China's recent achievements in space also highlights the importance of space in earthly affairs, as space is the location of much of the infrastructure that makes twenty-first century military and civilian operations possible.

We will return to some of these issues in later chapters as we make our way back to the present era of Mars exploration and the ways in which Mars can become a political battleground. For now, we have gotten only a taste of how and why robotic Mars exploration has become an endeavor that, while not necessarily the defining project of our era, is nonetheless considered important enough that multiple nations feel the need to pour large amounts of money into Mars missions. These examples speak to the political and cultural significance of Mars, and also to the general importance of space in our lives. Most of the time, the presence of space in our lives is unseen. Mars exploration and the politics surrounding it help to make our connection to space and space technologies visible.

### MARS: A HUMAN HISTORY

"The story which we are about to read has not been written by man, but, as the poet tells us, by the Creator himself; and therefore we can trust absolutely the truthfulness of the record. It is a plain unvarnished account in which man has no hand. . . ." Thus began the introduction to Henry Neville Hutchinson's 1890 book of popular geology, *The Autobiography of the Earth*. Within its pages, though Hutchinson's aim was to summarize and make understandable the evidence and theories of nineteenth-century geologists, he claimed, "The Earth is its own Biographer, and keeps its diary with the impartiality of a recording machine."[3] This book about Mars makes no such pretensions. Planets do preserve and erase elements of their own histories, but it is always humans who interpret and tell this history.

Mars is an object that has rarely spoken for itself, although it has at times been treated as animate. It has been with us from our earliest written records, and it will likely be with us until our end. But what it is, what it has been, and what it will be are not necessarily the same thing. The Mars I am interested in is Mars as we have known it, as we have described it, and as we have defined its importance to our world and our lives. This book is thus based, after a fashion, on the secondhand reports of the people who have mingled with Mars and who have made Mars central to their cultural projects. My guide in attempting to construct this portrait of Mars has been to attempt to follow the people who have cared about Mars, to ask why it was important to them and how it fit into their larger worldview, and to try to see what they saw.

In one poignant moment in Kim Stanley Robinson's 1992 novel *Red Mars*, Sax Russell, one of the scientists who has traveled to Mars as a member of the first human expedition, declares to his compatriots, "The beauty of Mars exists in the human mind. . . . Without the human presence it is just a collection of atoms, no different than any other random speck of matter in the universe." It is humans who understand Mars and give it meaning, he insists:

> All our centuries of looking up at the night sky and watching it wander through the stars. All those nights of watching it through the telescopes, looking at a tiny disk trying to see canals in the albedo changes. All those dumb sci-fi novels with their monsters and maidens and dying civilizations. And all the scientists who studied the data, or got us here. That's what makes Mars beautiful. Not the basalt and the oxides.[4]

Mars is presented as a natural object with a past that can be deciphered by scientists and their instruments, but with no history other than that which humans have given it. History, after all, is not the past but a way of talking about who we were. In this book I am just as concerned with who people thought they were—how they related to their universe—as I am in what ideas they had about Mars.

## WHO MADE MARS MATTER?

My task of getting beyond the basalt and the oxides hasn't always been easy. In every time period and culture I've examined, humans bestowed meaning, qualities, and characteristics upon Mars. But the planet's popularity wasn't always what it is today. For most of human history, Mars was only one member of a family of planets and stars whose significance came from their activities taken as a whole. In daily life, the Sun, Moon, and sometimes Venus were given special significance; these heavenly objects could be tied to regular natural and ritual cycles throughout the year. Because of its nearness to Earth during oppositions, Mars could become exceptionally bright and red and then seem to reverse course in the sky. Its brightness and its hue depended upon how close it was to the Sun during opposition, and early humans no doubt took note of these variations. Events like this gave Mars some dramatic significance. Venus—with its ability to appear either ahead of or behind the Sun as a morning or evening star, and to do so in regular cycles—was generally regarded with more interest. Mars, meanwhile, earned itself an impressive amount of distrust with its inconsistency as an occasional bad omen.

Within the past two hundred years, Mars has become central to the stories we tell about ourselves. It's an underdog story if ever there was one. And it is one intimately connected to our own changing conceptions of who and what we are. Our understanding of Mars has changed dramatically at different times in human history. Humans have seen and interacted with Mars in different ways at different times and places. This relationship with Mars has been defined by ideas and practices related to our understanding of the world and our place in it. Human definitions of Mars have been connected to definitions of ourselves, and understandings of our own significance. The outline of this story, as I will tell it, is as follows.

For the ancient Mayan, Chinese, and Babylonian cultures I address in chapter 1, Mars is part of an animate world in which humans and gods cooperate in the governance of human and natural affairs. While these three cultures approached astronomical and astrological knowledge systems in different ways, all three made the production of knowledge about the stars and planets a state-sponsored activity involving high-prestige experts. The cultivation and patronage of knowledge about the sky was a visible means by which rulers could demonstrate their commitment to the welfare of their subjects and the legitimacy of their reign. All three of these cultures ascribed meanings to the movements and appearance of Mars, and even tools for predicting its motion, and we will see these in some detail. We will attempt to understand these practices within the cosmologies in which they formed—cosmologies that concerned themselves as much with the role of humans in nature as with the planets' purposes. In this chapter we will also meet the ancient Greek and Roman philosophers who applied a new geometry to their world, and who reconceived the relationship between humans and planets to explain differences between themselves and their neighbors.

The goal of this book is not to present a chronology of how we came to understand the "real" Mars, or necessarily how modern science came to be—though the origins of modern science will be addressed. The history of Mars is not a continuous gathering of facts that at some point assembled themselves into a complete picture of the planet, nor is it the gradual stripping away of superstitious beliefs or the sudden awakening of a detached and objective view. My goal is to understand Mars in the context of the changing "world systems" that humans constructed. As we move into chapter 2 to view Mars through the eyes of medieval Europeans, we do so not to see how science has advanced, but to see how a new view of the world has reshaped how Mars and the other planets are seen. Here we will find Mars acting as a part of a living world machine, as an instrument through which God defines people and places, and also affects changes in a habitable world. We will see how and why these views developed through multicultural exchange between regions of the so-called Old World. And we will pay particular attention to those for whom knowledge of Mars was valuable—including not only natural philosophers and astrologers but doctors, necromancers, and religious thinkers. We will end the chapter with a discussion of two works of medieval literature that give us additional insight into how Mars, though not considered a physical world, was nonetheless visited.

In chapter 3 we examine how a changing conception of Earth and new technologies for observing the sky destabilized knowledge about Mars during the Renaissance. In the beginning of the time period covered in this chapter, Mars's significance was found in its effects on the human world—in its ability to bring things into existence and affect change in different parts of the world. We see how these ideas influenced an era of discovery and Eurocentric imperialism in the new Atlantic world. By the end of this period, Mars began to be seen as a world in its own right. However, the "Scientific Revolution" that is said to have occurred during this period did not occur quickly. Galileo's telescopic observations of the Moon and planets didn't immediately transform them into worlds; still other changes would make new worlds conceivable. These changes included changes to the way in which Earth itself was understood—changes that necessitated new frameworks for knowledge. Looking at an imagined version of Mars from this period, we see that old and new world systems overlapped, giving us a Mars that was in some ways similar to what we recognize today—a world with features of its own—but which was in other ways still rooted in a system that regarded Mars as part of an intelligent and purposeful world machine.

It is only in chapter 4 that we begin to see Mars as a "true" world with its own past, as we trace ideas about it from Johannes Kepler up through Percival Lowell. We see that this transformation was made possible not simply by better telescopes and mathematics, but by a newly defined family relationship between Earth and Mars, and the extension of a new tradition of terrestrial exploration and mapping. Rather than Mars being a force acting upon Earth, Mars and Earth both become planets subject to the same set of forces and made from the same original materials. New ways of producing and verifying knowledge emerge in this period, which sees the rise of professionalized science as well as the construction of boundaries separating various disciplines of knowledge. The story of Mars belonged to those during this period who wished to address questions about the processes that shape planets and their progress through stages of development and demise. These people included astronomers and, for a time, geologists. By the end of the nineteenth century the appearance or position of Mars in the sky no longer spoke to future events, but the features on its surface were understood to hold clues to the future of Earth as a living or dying world. As Mars became a distinctly Earthlike planet during this period, we also see an explosion of imagined Martian voyages and invasions during the eighteenth and nineteenth centuries.

The final two chapters of this book bring us into the more familiar territory of the space age. In chapter 5 we see how Mars exploration began in the context of the Cold War space race. The robotic exploration of Mars mobilized and reshaped the communities interested in the red planet, as spacecraft missions constituted a new form of "big science," with each mission involving hundreds if not thousands of participants. It also necessitated new relationships between the government, military, and universities. Twentieth- and twenty-first-century Mars explorations are conspicuous in that they take place on a world stage, with more immediate public access to images and data than in any previous period. During the Cold War, newspaper and television coverage brought Mars images into domestic spaces, along with images of Mars researchers and spacecraft engineers doing the human work of exploration. The images spoke not only to great feats of engineering, but to great communities of experts at the nation's beck and call. One very charismatic Mars explorer, Carl Sagan, became familiar to the public as the face of planetary science.

As the United States returned to Mars in the 1990s, the Internet and later social media came to define public engagement with the red planet. Twenty-first-century Mars exploration seems, at least on its surface, to be a continuation of Cold War exploration. And yet the Mars we now explore is very different from the one revealed by the Mariner and Viking missions of the 1960s and '70s. It is not simply a planet that has become better understood through new tools. New imaging systems with greater spatial and spectral resolution have revealed new features and mineralogical surprises that have transformed Mars yet again, into a world that may once have had a warm, wet past. Mars underwent a catastrophic transformation into an unlivable wasteland. And our culture has changed, too, in ways that go beyond the existence of smartphones and tablets. While the heavily cratered Mars of the early Mariner missions could be related to Cold War anxieties about atomic weapons and nuclear winter, this new Mars has become associated with one of the biggest existential crises humans face in the new millennium: climate change.

That we now see Mars as an inhospitable world has constrained the stories we tell about it. Mars fiction from the early space age on has presented a planet that must be endured and conquered, if not left to its own devices. Whether the stories we tell involve limited science missions to small research stations, or entire colonies living on Mars permanently, imagining a human presence on Mars raises immense engineering challenges. Nonetheless, for some a human future on Mars seems essential to the survival of the human species. As we will see, it often emerges within NASA as a long-term goal that will unite robotic and human exploration under one roof.

We will return to the present moment soon enough. For now, let's try to step back in time and see Mars through other eyes. Let's try to get a handle on the many lives Mars has lived in our human imagination.

# MARS AND THE COSMIC STATE

Who was the first person to notice Mars? When and where did people first start observing the motions of Mars and the other planets? What did Mars mean to them? These simple questions are impossible to answer, as these events likely happened in many places around the world long before people kept written records. We have records of such a small fraction—less than 2 percent—of the three-hundred-thousand-year existence of our species, *Homo sapiens*, and this moment is certainly buried unreachably deep in our past. But we can be sure of at least one thing: no matter where people were, they experienced the sky. They experienced it in a way that few humans today can.

If you've ever been far from the city on a clear and cloudless night, then maybe you've seen the sky in its full glory—with between two and three thousand stars visible to the naked eye along with the glowing plane of our Milky Way galaxy, and the planets slowly making their way through them. Today we live largely cut off from this spectacle, with smog and light pollution obscuring our view of this perpetual cosmic drama, the unfolding of which once seemed intimately connected to questions of both natural and social order.[1] But if we want to see Mars as it appeared to ancient humans, access to a clear night sky is only half of our problem. We also have to put ourselves in their minds.

It's tempting to say that we know more than the ancients. We have seen the surface of the Sun, the landscapes of Mars, the atmospheres of Venus and even Jupiter. We've discovered moons around other planets, and have found icy ocean worlds among them. We've used modern physics to infer the existence of black holes, and have now even "seen" them with powerful radio telescopes. We know that stars are bright orbs powered by thermonuclear fusion. We know that within the observable universe, many billions of them (a septillion, in fact, with twenty-four zeros) are packed into 170 billion galaxies occupying an expanding universe with a radius of around 46 billion light years (each light year is nearly six trillion miles). We know, in other words, that the universe is vaster than we can ever truly comprehend, that the starlight we see is millions of years older than we are (billions, if we go beyond what we can see with our eyes), and that there are more stars than we can count.

We might believe that we know the "true" nature of our universe, the objects held within it, and the forces that shape it. But the fact is that we aren't really any smarter than our ancient counterparts. The worldview of modern science and the technological capabilities we've developed for bringing the questions of science to the most distant objects of our universe are powerful. We've used them to construct a 13.8-billion-year history of our universe and the 4.5-billion-year history of our solar system and our planet, Earth. These tools have allowed us to detect and measure particles and electromagnetic wavelengths that we cannot otherwise see or feel; and on the basis of what we've found, we have constructed stories about ourselves and our universe.

But we aren't so different from those who came before us. Another thing science tells us is that the human brain has changed very little, if at all, over these thousands of years. Scientists disagree about when humans became capable of more abstract ways of thinking, but cave paintings suggest that it happened more than 150,000 years ago.[2] The humans who developed early astronomical knowledge were fully modern from a biological and cognitive perspective. We know that prehistoric and ancient people paid very close attention to the world around them and constructed complex worldviews to make sense of their universe(s).[3] They did not live in intellectual darkness any more than we do today. They asked questions. They solved the problems that were important to them. They learned about themselves and their world: the cycles and forces they saw or felt at work in ordering their existence. Myth helped them to see beyond what was apparent in their world, and to create holistic knowledge systems—to make connections that weren't otherwise obvious and create complete, coherent, and comprehensible worlds.

We know that early humans observed the night sky and saw patterns. There exist calendars of lunar cycles carved into bone as much as thirty-four thousand years old.[4] We also know that some of the most famous megalithic structures in the world, such as Stonehenge in England (ca. 3000 BCE) and Nabta Playa in southern Egypt (5000–4000 BCE), were used to unite the social lives of seminomadic groups of people with sky events. While most archaeologists no longer consider these sites to have been "ancient observatories" for predicting eclipses or the movements of the planets, the evidence remains convincing that ancient peoples gathered, performed rites, and worshipped in these spaces designed to celebrate celestial motions and alignments.[5] Anthony Aveni suggests that we consider these "sacred" observatories: "consecrated space[s] for watching the sky" in which "cosmic encounters were celebrated because they served to call people together to conduct rites to their gods."[6] Rather than attempt to make these observers conform to our idea of what an astronomer is or does, we can think of them as "sky-watchers" whose beliefs and practices varied across time and place. The truth is that we know very little about what preliterate societies knew or believed. But they left behind ample evidence of their attention to the movements of the Sun and the phases of the Moon. And we can be sure that whatever questions they asked of the heavens were very different from those that motivate space exploration today.

In reality, the difference between ancient and modern knowledge systems is more qualitative than quantitative; it is not about how much is known, but about what questions are important and about the acceptable ways of asking and answering those questions. And while we may not easily be able to slip between our modern worldview and those of others, we can nonetheless attempt to do so by asking not what ancient people knew about the world, but what their questions were when they looked at it. If we do this in the case of Mars, examining a few of the earliest known examples from around the world, we can see how sky knowledge was considered important to the functioning of the state—whether it was astrological knowledge in the service of good governance, or knowledge of bloodlines and relationships with the gods and other sky entities, which was used to legitimate power.

In each of the cultures we will look at here—the Maya, the Han Dynasty, and the Babylonians—the specific skywatching practices and associated belief systems varied. We will see that these ancient cultures resisted our modern tendency to compartmentalize different ways of knowing, understanding, and explaining the world (such as science, philosophy, and religion). After all, separating the study of the world from the search for meaning makes little sense when the world you inhabit is animate, intelligent, and full of purpose. As the examples we will look at illustrate, ancient practices of astronomy and astrology, understood in context, were not only ways of knowing the world, but ways of taking part in it.

### THE ANCIENT MAYA: REENACTING CREATION AND DECIPHERING THE MARS BEAST

Since I am writing this book in the Americas, it makes sense to begin this examination of Mars with ancient worldviews in the Americas. We begin with the ancestors of one of the largest and longest-surviving groups of Indigenous American peoples, the Maya. Mayan civilization emerged in the southern part of Mesoamerica, on Mexico's Yucatan Peninsula and in parts of Guatemala, Belize, and Honduras. Mesoamerican civilizations began to take root as early as 1500 BCE, with the first identifiably Mayan settlements appearing between the third and first centuries BCE.[7] The Maya lived in a universe in which everything was "alive with spiritual power," and in which little was truly inanimate. It's not surprising, then, that their depictions of Mars include what anthropologists have termed a "Mars Beast."[8]

The gods and demigods the Maya depicted in their art fall into four categories: worldly phenomena, anthropomorphs, zoomorphs, and animals. In the Dresden Codex, the "Mars Beast" appears as a zoomorph. It is not identifiable as any one animal, but is a chimera of various animal features—most noticeably, the cloven hooves of a deer or peccary (both animals are featured in Maya myth and astrology). The beast hangs upside down from segments of a sky band, representing the ecliptic (the Sun's annual path through the stars) and the constellations within (the Maya zodiac contained thirteen constellations). What the beast seems to be doing, confirmed by the accompanying text, is becoming brighter, showing more of itself (plate 1).

We might be tempted to interpret the Mars Beast as a bad omen. As we'll see later in this chapter, a lot of ancient cultures did associate Mars with death, war, and other forms of social and political turmoil. We can speculate as to why this was the case. One recent book about Mars accounts for its effects on the minds of prehistoric and ancient observers by invoking its "blood red" appearance, and describes its occasional seemingly irregular motion in the sky as being like that of a wounded animal wandering among the stars. The authors continue that "the pulsing light of Mars, mimicking the rise and fall of human emotions, the pulsating beat of the human heart, and the color of blood, must have taken its place in [early peoples'] myths."[9] Mars can indeed be noticeably bright and red, especially when it comes near to Earth during an especially close opposition. The outer planets beyond Mars have these moments, too, and all of them seem to wander when Earth passes them in their orbits, but none as noticeably as Mars. The distance between Earth and Mars over the course of each two-year cycle differs so much that the brightness of Mars at opposition can be fifty times greater than at other times. Its color can resemble wheat or a burning ember. It can seem not only alive, but impetuous, perhaps sinister.[10] Unlike most other things in the sky, it is not constant, or at least doesn't seem to be.

But the Maya don't seem to have viewed Mars in this way. Anthropologists today believe that the Mars Beast was not malign, but a signal of seasonal change. They suggest that the beast's behavior in the Dresden Codex, peeking out from the constellations of the sky band and showing more of itself, corresponds to what a skywatcher would witness as Mars comes nearest to Earth during opposition, entering retrograde, becoming brighter and redder.[11] The tables in which the beast appears, known as "water tables," correspond to the tropical growing seasons. Unlike in the mid-latitudes, agriculture in the topics is tied to a cycle of rainy seasons that don't precisely correspond to the calendar year. But the Maya seem to have found a connection between the motion of Mars and the seasons. The tables appear to relate to Mars's synodic period (the time it takes the planet to return to the same position relative to the Sun as viewed from Earth) and its period of retrograde. Here the Maya discovered "a pair of time cycles that . . . accurately described Mars's motion . . . [and] married it to other cosmic and terrestrial concerns."[12] The Mars table made it possible "for the ancient Maya to make a certain kind of prediction about the apparently erratic behavior of Mars that had both direct meaning and practical function for them."[13] This could have been valuable astronomical knowledge to the Maya, whose burgeoning population of two million was made possible by a vast agricultural system.

Sky knowledge was also connected to the social and political structure of Mayan community life. The Maya of the "Classic" period, from the third to the tenth century CE, occupied vast territory, but they were not an empire. Mayaland consisted of at least twenty independent city-states, each with its own capital and ruling family. Every city had a stone-built palace in which the king, his family, and his court and servants lived. Within the court were royal priests who performed rituals and ceremonies that maintained the connection between the king, his royal bloodline, and the ancestor-gods of the Mayan pantheon. These ceremonies marked the king as the descendent of the gods as well as the human interlocutor between the gods and the Maya people, ensuring the prosperity and health of his city.[14] The architecture of the palaces and ceremonial spaces commissioned by the ruling elite conveyed religious and political information, and served as "great stage fronts for the rituals vital to the sustenance of society as a whole."[15] This was accomplished in part by the inscriptions and images that adorned these spaces. But it was also the result of the cooperation of the planets and stars, which on some nights could be seen reenacting important historical cosmic events in their movements through the night sky.

Predicting on what nights a ceremony including celestial movements could be performed, not to mention orienting ceremonial spaces to best showcase these movements, was no small feat. Palaces employed royal scribes (*aj tz'ih*) and priest-scribes (*aj k'uhuun*), the latter of whom likely acted as court historians, genealogists, mathematicians, astronomer-astrologers, and masters of ritual.[16] These scribes prepared and tracked the ritual calendar, and recorded the rituals performed and the myths reenacted to legitimate the king's divine rulership. Most commonly, the king was connected to one of the twin hero gods of the Mayan creation story, and rituals of legitimation included reenacting the god's exploits, including a trip through the underworld into the sky.

The creation story is central to ancient Mayan cosmology, as it explains the order of the universe and the human relationship to the gods. We know this story today mainly thanks to a sixteenth-century text called the *Popol Vuh*, a Spanish transcription of a text containing the ancestral knowledge of the K'iche' Maya of the Guatemalan highlands. We know that portions of the text were most likely meant to be performed in public reenactment over the course of multiple nights, and that these performances were part of the annual rituals that legitimated the king's divine rule. Depictions of these reenactments from as long ago as the sixth century CE, along with the dates on which they were performed and the names and lineages of the kings who performed them, can be found on monuments throughout Mayaland.[17]

The *Popol Vuh* describes a universe inhabited by a pantheon of gods, demigods, and supernatural creatures related to the Mayan experience of the world—the plants they cultivated, the animals they hunted or feared, and the sights and sounds of life in Mesoamerican farming communities. The main story of the *Popol Vuh* is the creation of the universe, or "sky-earth." It begins with the creation of the primordial world from the sea and the night: "The broad sky is all alone. The face of the earth is not yet here. The expanse of sea is all alone, along with the womb of the sky."[18] Chief among the gods of the first creation are Xpiyakok and

Xmukane, though they need the assistance of a collection of gods to enact the creation. The council of gods deliberates, plans, and comes to consensus.[19] As one, "they reached an accord, braiding together their words and their thoughts," and they give birth to the world and humanity.[20] By saying the word "Earth," they immediately create Earth. They call forth the mountains, valleys, and waterways, separating the waters from the land.

The creation narrative is interrupted by the exploits of gods and demigods in the prehuman world, and these stories establish the Maya cosmos as consisting of Earth, sky, and underworld. The two main protagonists of these stories are the twins Hunahpu and Xbalanque. The culmination of their adventures is their defeat and humbling of the gods of death in Xibalba, the underworld. After bringing the gods of death in line, the two brothers ascend into the sky, one as the Sun and the other as the Moon: "Then the twins rose together as the central lights of the world. They ascended straight into the sky: one arose as the sun, the other arose as the moon. And the womb of the sky was filled with light, and also the face of the earth. They dwelled there, in the luminous sky."[21] This story not only explained the significance of the Sun and Moon in Mayan cosmology, but defined a path to the sky through the underworld that Maya rulers had to reenact ceremonially during their lives, and ultimately one last time when they died.

Only after the two brothers have ascended into the sky does the final creation of humans commence. The council of gods convenes to discuss the creation of humanity. They create four men from maize. These humans are perfect in every way. In fact, they are too perfect. Their sight is too keen and their intellect too sharp. They are able to see through and understand all of creation. As one of the gods explains, "It is not good what we discovered. Their deeds could rival ours. Their knowledge reaches far. They could grasp everything."[22] The gods correct their mistake by taking back their knowledge of the world and blurring their vision. Once the gods are happy with their revised creation, they create the first four women as wives for the four men, and these become the founders of the Maya people: the "bloodletters and sacrificers" whom the gods have created to worship and pay tribute to them.[23] Thus it is established that humans will have imperfect knowledge of their world, that their purpose will be to serve and pay tribute to the gods, and that the working of the world will depend on a relationship between humans and the gods.

The *Popol Vuh* contains the core of Maya cosmology, but it is in other sources that we find more direct links to the sky. Some inscriptions related to the creation story found on stone monuments in ancient Mayan cities contain astronomical variations that help us understand how reenactments of the creation might have incorporated the movements of the stars and planets. At one site in Guatemala, anthropologists found the description of three stones being placed on Earth to allow the first fire of creation to be lit, and the sky to be lifted from the primordial sea.[24] In some variations of this story, it is the maize god who places the stones so that the sky can be raised. But in the version found at this Guatemalan site, the sky is still lying on the ground. The three hearthstones, laid on the sky, are understood to be three stars that form a triangular pattern in the constellation Orion: stars we know today as Alnitak, Saiph, and Rigel.

Once the sky was lifted, mirror images of the triangular pattern on the earth and in the sky joined the two realms together. The great Wakah-Chan (World Tree) then lifted the sky into place, ordered the cosmos (placing the constellations in the zodiac and the Sun and planets on the ecliptic), and began the turning of the heavens, thus initiating time and space. Anthropologists today believe that this World Tree refers to the Milky Way, alternately known as Sak Be (White Road) and Xibal Be (Road of Awe). Noting these associations, the archaeologist Linda Schele argued that "the final events of the Creation were all played out in the sky" as these celestial objects moved through the sky and the Milky Way formed a cross with the ecliptic on the night of the final creation—which was, by Mayan calculations, August 13, 3114 BCE.[25]

The sky replayed this creation narrative over the course of its annual motion. Maya priests and scribes no doubt spent a great deal of time establishing the methods of predicting on which nights the creation story should be reenacted. It had to be a night on which participants and viewers would see the World Tree unite the underworld with the middleworld (Earth) and overworld (sky), allowing the king to pass into these other worlds and, in so doing, become the maize god for his people. Opening such a "portal of communication between the inhabitants of the sky and the earth" was vital for maintaining a Mayan city's religious charter;[26] the associated calendrical practices for ensuring the planets' cooperation helped to maintain "the rights and responsibilities of the ruling elite," and were patronized by the ruling class for this purpose.[27]

Mars played its part in Maya rituals. In the Maya city of Palenque in southern Mexico, the ruling family had built a triad of temples to replicate the three stones of creation, symbolically locating the city at the center of the Mayan cosmos. Using modern planetarium software programs to replay the night sky on different historic occasions, anthropologists have recreated the scene in Palenque when Chan Bahlum ascended to the throne on July 18, 690, a night when three creator gods were seen to reunite over the city. After sundown, three planets lined up from east to west in the constellation Scorpio: Saturn, Jupiter, and Mars. A bright and gibbous Moon joined them, and all four bright objects passed over the southern sky. As one anthropologist describes the sight:

> One can imagine a crowd assembled in the open plaza facing Chan Bahlum's temples, watching the reunited divine trio who had given birth to their ancient ancestors as they disappeared over the Temple of the Inscriptions, housing the tomb wherein Chan Bahlum's father recently had been laid to rest—a symbolic affirmation that the celestial power with which the father was once endowed would pass on to his son.[28]

The historian of science Gerardo Aldana suggests that Mars on that night was associated with Itzamnaaj, one of the original gods of the creation (hence one of the oldest) and one of the gods' most powerful priests. Mars is one of the slowest planets in the night sky; the Maya approximated its synodic period as 780 days. But at times when Venus is not in view, it can be the brightest planet in the sky during its opposition. As "Mars appears as the slowest yet most potent of the celestial bodies, Mars becomes a suitable celestial realization of Itzamnaaj."[29]

## ANCIENT CHINA: THE HEAVENLY MANDATE AND THE SPARKLING DELUDER

In the third century BCE, the Qin dynasty initiated China's imperial period by consolidating the independent states that had grown up around agricultural communities in the previous century. This first dynasty ended in rebellion and collapse, and was replaced by the Han dynasty in 202 BCE, under the rule of former rebel chieftain Liu Bang, who became the first Han emperor, Emperor Gao. While the Qin dynasty had tried and failed to maintain power through bureaucratically imposed draconian laws, hard labor, and taxation, Emperor Gao and his descendants in the House of Liu ruled over a less centralized collection of feudal states, and spent much of their rule struggling "to define the proper relationship between the imperial government and the neofeudal kingdoms."[30] Astronomy would be central to these efforts.

Long before the rise of the Qin dynasty, Chinese culture had developed the notion of the "mandate of heaven"—the idea that only a family of great virtue and spiritual power could rule China. The rulers from such a family would serve as heaven's ministers.[31] By "heaven," what was meant was the natural order: in ancient Chinese cosmology, the universe was a mirror of human society in which human bodies, communities, and the heavens were "mutually resonant systems of which the emperor was the indispensable mediator."[32] For the Han dynasty, the mandate could not be taken for granted. Liu Bang had come from common beginnings. Once in power, the dynasty had to work to maintain the belief that the emperor held the mandate through ritual and architecture, offering prayers and sacrifices to heaven in specially designed temples where the relationship was forged and maintained.[33] The Han dynasty took special care not only to observe rituals (and to be witnessed doing so), but to patronize great astronomical reform projects that would help it better follow the way of heaven.[34]

The emperor maintained harmony between the affairs of humans and the workings of the cosmos primarily through the service of scribes (*shih*). Like the Mayan scribes, these scribes and the grand scribe (*T'ai shih*) were specialists of high social rank employed by the ruling elite to seek out and decipher omens.[35] Astrology was one of the primary means through which scribes provided counsel to their emperor, and they used records of observations and correspondences compiled over centuries to decipher *tian xiang* (heavenly symbols) through a practice called *tian wen* (heavenly writing).[36] As one Chinese source explained, in tian wen the astrologer follows the movements of the planets, Sun, and Moon through the constellations "so as to set in order the phenomena [that show] good or evil fortune, [all of which] the sage ruler uses to calibrate his government."[37] A good astrologer identified significant celestial events for his emperor, and connected them to corresponding earthly events.

In the early years of the Han dynasty, Liu An, a member of the ruling family and the king of the feudal state of Huainan, wrote a book on governance that synthesized existing traditional knowledge and schools of thought on the nature of the cosmos and the place of humans within it. Primarily Daoist in its approach, Liu An's *Huainanzi* made no distinction between the workings of nature and the affairs of humans, accentuating instead the harmony that united them. Human actions were constrained by cosmological principles. Only a ruler who could make his decisions and actions conform to the movements of heaven and the workings of Earth could hope to control human affairs.[38] The consequences of bad rule would be felt in nature: cruelty would result in whirlwinds, unjust laws in plagues and pestilence, persecution in drought, and unruliness in torrential rains.[39] Only good and just decisions maintained harmony. Therefore, a good king had to acquire "penetrating insight" and learn all nuances of the natural order.[40]

Three chapters of the *Huainanzi* cover cosmological issues including astronomy. astrology. and the ritual calendar. These chapters present a worldview shared by the majority of early Han intellectuals which remained influential for centuries, and which likely had developed over centuries before the Han dynasty. On a collection of etched Shang dynasty (1559–1046 BCE) "oracle bones," made mainly from smoothed and polished tortoise shells and ox scapulae and used for divination, we find evidence that the people of early neolithic China watched the stars with interest. They recorded star names in the bones that would later be categorized into the twenty-eight mansions (constellations) through which the Sun, Moon, and planets moved. In the fifth century BCE, during the Warring States period, handbooks not so different from the *Huainanzi* began to compile the ancient annals of the universe. By the Qin dynasty, cosmological speculation had become a state-sponsored activity: astrology, as it served greater bureaucratic purposes, was almost a civil service. The *Huainanzi* was written in conversation with these other more ancient texts and cosmological traditions.

The narrative presented in the *Huainanzi* is used to tell us the sources of order and chaos in the universe, and to give us a description and explanation of the universe's structure. The narrative presented is also then subjected to analysis, as Liu An devotes a good deal of attention to drawing out truths from the narrative and synthesizing them with astronomical and astrological knowledge. We find this primarily in chapter 3 of the *Huainanzi*, "Tianwenxun" (The patterns of heaven). This chapter is a treatise on the interconnectedness of all things in the universe, and on the cosmic cycles and correlations that govern all things on Earth. The purpose of the treatise is to instruct the reader, first, in how to observe the heavens and make sense of observations; and second, in how to follow the portents seen in the sky, to avoid going against the cosmic order. "Tianwenxun" does not actually teach astrology in detail, as it assumes the presence of a court astrologer for this purpose; however, a close reading of the chapter would leave one generally conversant in astrology and able to follow the advice of an expert.[41]

The chapter begins with the origin of the cosmos, presented in a poem that is itself based on an earlier poem, "Tianwen" (Heavenly questions) by Qu Yuan. Liu An's revisions and additions attempt to answer some of the questions the earlier poet raised. In the account, all begins in unformed chaos until an inherent dualism separates the basic materials of the universe. After the original universe is formed, an epic battle between Gong Gong and Zhuan Xu results in the tilting of the heavens, thus explaining why the Sun's apparent path around Earth does not follow the celestial equator. This battle marks the beginning of linear time and the beginning of the human era of thearchs (the era of heavenly monarchs who worked to harmonize the now separated sky and Earth).[42]

Achieving this harmony required some understanding of the structure of the heavens, including its height and its divisions. The book provides information about the five planets, which are called the five stars or, in an allusion to weaving, the five wefts. It is here that we learn about *ying huo* (the Sparkling Deluder, Mars)[43] and its associations:

> The South is Fire. Its god is Yan Di [Flame Emperor].
> His assistant is Zhu Ming.
> He grasps the balance-beam and governs summer.
> His spirit is Sparkling Deluder [Mars].
> His animal is the vermilion bird [a mythical bird similar to the phoenix].
> His musical note is zhi;
> His days [heavenly stems] are bing and ding.[44]

Like all of the planets, Mars is given a phase (fire), a direction (south), a color (red), a season (summer), a musical note (*zhi*), and two heavenly stems (*bing* and *ding*). The southern sky, where Mars is said to have special influence, is guarded by the phoenixlike vermilion bird, which itself resembles a pheasant with flame-red feathers (figure 3).

We also learn about the motions of each planet. For Mars, we are told general details about its motion through the sky, and are also warned about its negative connotations and associations with those who do not act to maintain the harmony of nature (those who do not follow the way):

> Mars normally enters the asterism Grand Enclosure in the tenth month.
> [The corresponding state thereupon] comes under its control.
> Then it emerges, passing through the lunar lodges in turn.
> (Mars) governs states that lack the Way,

**FIGURE 3** The vermilion bird depicted on a tile from Han dynasty China. Gift of Charles Lang Freer. Smithsonian National Museum of Asian Art.

> Causing disorder, robbery, sickness, mourning, famine, and warfare.
> When its leavings and enterings . . . are irregular, there is disputation and change.
> Its color is sometimes visible and sometimes unnoticeable.[45]

Mars, the Sparkling Deluder, was an ill omen. Its association with the south, mentioned above, meant that it could be an omen of military conflict. It's associations with fire and summer could also make it an omen of drought.

### ANCIENT MESOPOTAMIA: MARS AND THE NEGOTIATION OF BEST FUTURES

Now we travel to the site historically associated with the origins of Western science, once thought to be the "cradle of civilization"—ancient Mesopotamia. As the above examples of the ancient civilizations of

Mesoamerica and China illustrate, there were in fact multiple regions around the world where complex agricultural civilizations originated, and we no longer privilege Mesopotamia as we once did. But we remain interested in what happened in Mesopotamia for at least two reasons. First, the cosmology developed there by the Babylonian and Assyrian empires endured for centuries, leaving long records of observations and the development of mathematical tools for predicting sky events. Second, these records and pieces of the cosmology were adopted by the Greeks and used in their astronomy-astrology practices, thus influencing the actors in our next two chapters. Babylonian astronomical and astrological knowledge possibly even traveled as far as ancient India, and may have influenced early native work there.

The inhabitants of ancient Mesopotamia would not have referred to themselves as "Mesopotamians." The Greeks gave the region its name, marking it as the "land between rivers," a reference to the Tigris and Euphrates river system that surrounded the region and facilitated the rise of irrigated agriculture and the cultivation of grain beginning around 5000 BCE in Sumer (present-day Iraq). As the region developed, independent city-states ultimately became an empire with vast trading routes extending throughout the Middle East and to India. Early forms of pictographic writing emerged here in the middle of the fourth millennium BCE, followed around 3000 BCE by the syllabic script, cuneiform, which involved symbols pressed with reeds into clay tablets. Due to the relative abundance of clay in the region, and the development of literacy, we know a good deal about the Mesopotamian worldview, fragmentary though these records may be.

The period we are most interested in begins around 1900 BCE, with the rise of the Babylonian empire. Babylonian cosmology was directly related to social structure; like the Maya and the ancient Chinese, Babylonians believed they lived in a cosmic state. Their human world was connected to the celestial world of their gods as one entity. The decisions of the gods and goddesses—who lived in a divine council—were transmitted through the stars to be read by the astrologers who in turn advised the king. Economic and political stability were ensured by pleasing the gods.[46] The gods were concerned with human well-being, and knowledgeable astronomer-astrologer priests knew not only how to read the heavens, but how to use a form of apotropaic magic (ritual practices for avoiding evil) they termed *namburbu* to persuade the gods not to follow through on their bad omens.[47]

In the next chapter we will encounter a form of astrology that is based on planetary influence, in which the planets themselves are seen as forces of change in the world. Babylonian astrologers understood the planets very differently, and saw no direct causal relationships between the movements of the planets and occurrences on Earth. Nor did they view the stars or planets as objects of worship. Instead, much as in Chinese cosmology, the heavens contained messages from the gods that foretold what might happen if appropriate action wasn't taken.[48] It is also significant that, while the Babylonians did develop a method for predicting the motions of the planets, they did this through purely mathematical instruments, without a geometrical (spherical) cosmological model.[49] At no point do the Babylonians seem to have been concerned with the structure of the heavens; they cared only about its function in the reciprocal relationship between gods and humans.

Examining the case of Babylonian astrology, the historian of science Noel Swerdlow argued that "in the belief in omens lies the birth of science."[50] The historian of astrology Nicholas Campion agrees, calling what happened in Mesopotamia "a small scientific revolution."[51] What Swerdlow and Campion mean by this is that the belief that the sky held warnings of what was to come led to systematic observation, recording, and eventual prediction of the positions of the stars and planets in the sky. The practices required for this task led, as in our previous two cases, to the establishment of a class of experts who worked for centuries to create an extensive body of knowledge about the heavens and the phenomena to be found therein.[52] But before we wade too deeply into a description of the Babylonian astronomer-astrologers, it will be helpful to know a little more about the cosmos as they understood it. As we did with the Maya and the ancient Chinese, we should understand their view of creation.

The Babylonian creation myth is the *Enuma Elish* (When on high). We don't really know when the pantheon of Babylonian gods emerged, or much at all about the oral tradition in which their creation narrative evolved, but the clay tablets that form the *Enuma Elish* hold the most complete version of this story. Perhaps representing the region's dependence on the river system, the Babylonian world begins with water. In the beginning, this water is chaos—swirling and undifferentiated. The first move of creation is the division of the fresh water from the brackish. The sweet water becomes the god Apsu, while the salt water becomes the goddess Tiamat. These two original deities become parents to the younger gods of the Babylonian pantheon, and intergenerational conflict follows. Ultimately, the two elder gods are slain by their children. The hero Marduk, patron god of Babylonia, creates the cosmos from the body of Tiamat and assigns the other gods their roles in maintaining the new universe. The god Ea creates the first humans from clay mixed with the remains of Marduk's enemies. These humans, like the first Maya, were bound to the gods in their service and were given a role to play in helping the gods maintain the order of the universe and fight back chaos.[53]

The covenant between the gods and the Babylonian people is relatively straightforward: the gods will rule the heavens, and the humans will rule Earth. But it would be a mistake to think of heaven and Earth as separate realms, or to think of the gods as all-powerful deities. The Babylonian gods ruled all of nature, and at the same time they were nature; they bestowed order, but were at the same time subject to that order. This is established in the *Enuma Elish* when Marduk creates a system through which the gods and humans will communicate. He orders the Moon to create the calendar, and creates a sky that moves with observable regularity. Marduk establishes the star Nibiru to shepherd the planets, which in Sumerian were known as *udu idim mes* (the wild sheep).[54]

When anything in the sky deviated from regular motion, or when something new appeared in the sky (including atmospheric phenomena), humans would know that this was a portent or omen sent from the gods to warn them of a coming event.[55] In this system, the planets were seen as both orderly and disorderly. Their motions were governed by the same intelligent forces of nature as those of the Sun and Moon, but were also less predictable. In these deviations from order the Babylonians found messages from their gods.[56] Though the planets themselves were not gods, and were not worshipped as such, each planet was associated with a god from the Babylonian pantheon (when these associations were established is not known). Venus, to which the Babylonians paid special attention, was associated with their god Ishtar. Mars was associated with Nergal, a god of the underworld—a fitting association for a planet whose omens tended to presage fire, plague, war, and other misfortunes.

Babylonian astrology provided a means of communicating with the gods, through which kings could negotiate the futures of their kingdoms. It was thus primarily focused on questions related to the welfare of the state, politics, and economics, as they manifested in weather cycles, harvests, political rivalries and alliances. Such negotiations required astrologers who knew how to read the sky, knew which events were associated with sky events, and knew how to speak back. The Babylonian astrologers were an elite, scholarly caste. Their practices drew upon centuries of diaries written by their predecessors that contained detailed records of occurrences in the sky as they corresponded to detailed accounts of major earthly events, all compiled with the intent of understanding the reciprocal relationship between heaven and Earth.[57] The scribes were diviners who knew how to speak to the sky. When they looked to it for messages, they also called upon it to open channels of communication. As one prayer of divination began, "Stand by me, O Gods of the Night! Heed my words, O Gods of destinies."[58]

We have some idea of what omens the astrologers looked for and recorded. A set of seventy tablets known as the *Enuma Anu Enlil* (When Anu and Enlil) contains around seven thousand omens. As we might expect, the Sun, Moon, and Venus get the most attention. However, four of these tablets are believed to be dedicated to Mars. We also have the so-called "Astronomical Diaries" tablets, containing the *nasaru sa gine* ("regular watching" reports) prepared by the astrologer scribes.[59] And we have tablets that were essentially letters written between the astrologers and their royal patrons, explaining the omens they witnessed. In these, we find the if-then structure of the Babylonian omen system: if thing A is seen in the sky, then thing B will follow.

What do the tablets say about Mars? One thing we learn from Babylonian astrology is that the meaning of Mars can vary with the season and the planet's location. Under some conditions, the appearance of Mars can mean that "the cemetery of warriors will enlarge." In others, it can mean the death of a king or political instability.[60] We also see that sometimes the omen is influenced by the planet's movement through the constellations and where it begins its retrograde motion. When retrograde begins in Scorpio, "the king should not go outdoors on an evil day." When retrograde begins at Leo, the king's reign may be at its end.[61] If Mars stops in a constellation and becomes bright, as it might during opposition, this can also be an omen. Depending on the constellation, it could mean devastation or prosperity.[62] Almost all of these omens sound pretty bad, but one would want to be able to spot the good omens, too. A good king would want to avoid as much misfortune as possible, and to accentuate whatever good might come. Only a properly trained astrologer could decipher these signs.

Fortunately, the astrologers' expertise included the rituals the king could perform to communicate back to the council of gods and ask respectfully for the best possible future.[63] The *namburbu* mentioned above contained detailed instructions on what a king should do to alter the future the omen predicts—to enhance favorable outcomes and avoid undesirable ones.[64] Mesopotamian astrology was performed in service of the state, in order to preserve stability and order. While kings did not rule by divine right, the king who performed his ritual duties to negotiate the best future for his people was seen to be upholding the human obligation to rule alongside the gods. Mars, then, was just one part of this complex system.

## ANCIENT GREECE: THE GEOMETRIC AND GEOCENTRIC WORLDVIEW

Finally, we have come to ancient Greece, the time and location most associated today with the birth of science. As our previous examples should make clear, this is not a straightforward claim. Elements of science, including observational methods, mathematical tools, and predictive methods were in use in locations around the world long before the rise of Greek philosophy. Furthermore, if we found elements of myth or the supernatural at work in the previous three cultures examined that seem to disqualify these worldviews as scientific, or that illustrate that those cultures were using "science" for nonscientific purposes, then we will encounter much the same problem in Greece (and in medieval and early modern Europe, for that matter). Nevertheless, the texts that would later be considered the starting point for conversations about nature and the order of the cosmos, and with which scholars would continue to engage well into what we refer to as the "Scientific Revolution," originated in Greece. And so it is important that we spend some time understanding this worldview on its own terms, while being careful not to privilege it as being somehow superior to the other worldviews we've examined up to this point. Each worldview was devised in specific cultural contexts and was constructed from the observations, questions, and priorities most important in that context. The Greek example is no different; it did not benefit from any special access to the natural world or its secrets. Still, our focus here will be on what was unique about the Greek approach to the world, and how it understood the place of Mars in that world.

What was unique about Greece? The Greeks benefited from their adoption of a modified Phoenician alphabetic writing system around 800 BCE, which allowed the subsequent expansion of literacy over the next few centuries. Reading and writing were no longer limited to the highly trained elite, but could be learned relatively quickly by anyone. Historians consider ancient Greece beginning in the sixth or fifth century BCE to have been the world's first widely literate culture. The Greeks, of course, like all other cultures, had a preliterate oral history tradition that informed what we find in their writings. They also had a very well established mythology, including a pantheon of gods and goddesses, and epics concerning the exploits of heroic humans and demigods. Greek poets like Homer and Hesiod, with their descriptions of the gods, their origins, and their interventions in earthly life, provide us with a fairly complete understanding of Greek mythology.

Greek mythology is so well known that it is hardly worth examining it in detail here. Even today, reimagined versions of the Greek and Roman gods appear as characters in modern popular culture. We will simply note that the planet Mars was associated with the Greek god Ares, the god of war, and was often considered an ill omen. We can also note that the word "planet," which we've been using for several pages now without comment, originated from the Greek word for "wanderer," binding the planets' identities to their apparent motion through the fixed stars. But while the word "planet" originated there, Western planetary knowledge did not. The ancient Greeks adopted Mesopotamian astrological ideas, among others, and adapted them to their own pantheon. They likewise made use of Babylonian records of astronomical observations. And they spread this knowledge. Perhaps because of the increase in literacy, the proliferation of schools, and other social changes that occurred in Greece, astrological knowledge, while still esoteric, was no longer considered the domain of kings and royal astrologers. It was instead understood to be a means through which any individual could divine their fate. There was also a shift away from understanding the planetary motions as messages from the gods, and toward considering the planets as causes or influences of earthly events.

One reason why astrology ceased to be a conversation with the gods was the new philosophy. The Greek philosophers moved cosmology at least a half step away from personification or deification of the forces of nature. Their view of the universe was more mechanical, and their approach to answering questions about the world was to define their own rules of logic and then apply them to large categories of natural phenomena. They tended to find the causes of phenomena not in external interventions, but in the nature of things themselves.[65] They developed physical explanations for why things happened in the sky, even as their physics continued to require what we would today consider supernatural or at least extranatural forces to perpetuate motion in the heavens. Quite a few philosophies of nature proliferated in ancient Greece. For our purposes, we will be looking at three thinkers whose ideas became influential to the medieval Islamic, Jewish, and Christian natural philosophers we will meet in the next chapter.

For cosmology, we turn to Plato. Plato's main cosmological work, the *Timaeus*, offers a creation story based on what was in his time the best available scientific knowledge (as anachronistic as this may sound). Far from being divinely inspired, Plato's creation story is, as he describes it, only a "likely story" (*eikôs muthos*) of how the cosmos came to be, based on what was known of the world. Plato begins with the observation that the world is rational, and that it thus must be the creation of a rational mind or intellect. Rather than adopt one of the existing anthropomorphic gods of the Greek pantheon, Plato posits instead that the world is the creation of a demiurge, or divine craftsman. The craftsman's design is purposeful, made of parts that work together toward beneficent ends. Plato's world is not perfect, however. The divine craftsman may have been inspired by perfection, but the primordial elements used to build the world have their limitations. The demiurge managed to order chaos in building the world, but the materials of the world refuse to remain stable.[66] Everything in the world is thus an imperfect version of its ideal form, and the world remains a place of change and corruption.

Because the divine craftsman was rational and good, so too is the universe. It is a living thing with a soul of its own, and this world soul penetrates the entire universe and is responsible for the motion of all of its parts. The stars and the celestial bodies, which move in circular motions around the central, spherical Earth, provide markers of time, day and night, the month, the year, and so on; and it is through their intelligent motions that time is brought into existence. The planets are divine, as gods, but not as anthropomorphic gods who intervene in human affairs; they are steadfast and seek never to deviate from their rational motion.[67] Human souls, in contrast to the world soul, are made from the imperfect residue of the universe. When the human soul is then placed within a physical body made from the imperfect elements, the soul becomes confused and unable to see or remember the order of the universe. Only through education and disciplined thought, through nurture of the intellect, can the original dignity of the soul be restored. This can be done by looking to the structure and functioning of the intelligent and purposeful world as an example.

The *Timaeus* does contain basic physical explanations for how and why things happen in the universe that are based in the earlier Pythagorean school of geometry and mathematics, as well as other precursors. However, the most influential system of physics was developed by Plato's student Aristotle, and so we turn our attention to him. We will address the four Aristotelian elements—earth, water, air, and fire—in the next chapter, as they pertain to thinking about change in the sublunar world and the role the heavens play in influencing that change. Here, we will note that Aristotle based his physics on observation of the world around him. As he saw objects move on Earth, he noticed that when left to their own devices, they tended to move either up or down. Fire and hot air tended to rise. Water (rain) tended to fall, as did earth if it was dropped from any height. To move in any direction other than up or down, unnatural force had to be applied. Heavenly objects, on the other hand, always moved in circles, and never up nor down. Thus, Aristotle concluded that the heavens must be made of different physical stuff than earthly nature: a fifth element called aether. While the four elements of the sublunar world had in their natures the tendencies to move up and down, the fifth element was only capable of perfect circular motion.

Aristotle created two distinct realms, the substances of which could not mix, and the natural motions of which were different. Below the Moon, the four elements made up everything. Above the Moon, all was made of aether. Nowhere within this system was there empty space, nor was there anything that lacked an essential nature. Aristotle's world was full, and, like the world of Plato's *Timaeus*, it was one of order and purpose. His physics reflect this in the idea that the nature of things determines their behavior; the order is built into the material fabric of a universe that is everywhere connected.[68] There remained the question of why the planets and stars continued to move. Where Plato had his demiurge, Aristotle introduced an unmoved mover. Each of the planets, including the Sun and Moon, was placed within a series of nested spheres; each sphere turned because of its love for the unmoved mover and its desire to imitate it.[69]

The final piece of the ancient Mediterranean worldview we will examine is the detailed arrangement of the heavens provided by the Greco-Roman astronomer Claudius Ptolemy. The culture of Hellenistic Greece and later Rome was markedly different from that of the Greece of Plato and Aristotle, as it was influenced greatly by increased contact with non-Greek cultures as a result of Alexander the Great's imperial ambitions. Alexander not only conquered the existing Greek city-states but waged military campaigns into the Middle East and north Africa to overthrow the Persian Empire. After he died in 323 BCE, Alexander's warring generals split his huge empire into several Hellenistic states dominated by Greek-speaking elites. When Rome rose to power in the second and first centuries BCE, it absorbed the former kingdoms, with the result that Greek was the official language of its eastern empire. Ptolemy wrote his works from the Roman Egyptian city of Alexandria, where he was associated with the city's museum and libraries. There he had access to Greek, Egyptian, and Babylonian knowledge. Ptolemy's planetary astronomy, expressed primarily in his *Almagest*, was perhaps the most mathematically sophisticated astronomical work the world had yet seen.[70]

Ptolemy's aim was to describe and predict the observed positions of the planets through combinations of perfect circular motions. As his starting point, he adopted a nested sphere model similar to Aristotle's. However, as any careful astronomical observer can see, the planets do not move around Earth as though attached to a fixed point on a sphere; as we know today, the planets orbit the Sun, not Earth, and Earth itself rotates on its axis as it orbits the Sun. To correct for these apparent discrepancies, Ptolemy introduced eccentric spheres, which orbited a point other than Earth's center. To explain retrograde motion, he introduced epicycles: smaller spheres orbiting points on the eccentric sphere. While Ptolemy's system was not perfect, it did succeed in making planetary astronomy predictive, and it preserved the quality of uniform circular motion. Thus, his mathematical model could be justified through existing arguments from Greek natural philosophy.[71]

Ptolemy's contributions didn't end with his predictive planetary astronomy. He had an idea of the value of this tool to the historical science of astrology. In addition to his *Almagest*, he also wrote a set of four books, traditionally called the *Tetrabiblos*, devoted to a systematic study of the effects of the planets on the sublunar world and the practice of astrology. These works contain arguments in favor of astrology, as well as an encyclopedic presentation of ancient astrological knowledge. As we will see in later chapters, this went beyond what we think of today as astrology; it was interested in how the planets affected all aspects of the world, including what people, animals, and things would be found in different regions. It reflected not only Greek and Roman knowledge about the world, but also prejudices about the peoples of other lands. Book 2, devoted to "mundane astrology," presents several characterizations we would recognize today as ethnic stereotypes, and relates them to the influence of the various planets and constellations associated with the different inhabited regions of the known world. The final two books address astrology as we recognize it today, in the form of individualized horoscopes.

Along with the other planets, Ptolemy's Mars had its role to play in determining the mental character and physical features of the people of the world, and in the properties of the lands they inhabited. But we will save a close examination of its properties and influence the next two chapters. As we will see, the ideas of Plato, Aristotle, and Ptolemy shaped centuries of scholarly discussion and debate in the Middle East and Europe. Plato's cosmology, including his introduction of the soul and the embodied soul's confusion, would become important in defining the purpose of natural philosophy in the early Middle Ages. Aristotle's physics—his notions of the four elements and the aether, of a universe composed of nested spheres, and of a prime mover—would shape discussions about the shape and purpose of the heavens as it was reconciled with the new monotheisms of Judaism, Christianity, and

Islam. Likewise, Ptolemy's synthesis of Greek, Babylonian, and Egyptian astronomy and astrology would become the basis for nearly all observational and predictive astronomy and natural philosophy in the West until the seventeenth century, and would inform debates about how God acted upon the world through natural means.

# MARS IN THE MEDIEVAL IMAGINATION

From the Middle Ages into the Renaissance, Europeans believed not only that the planets affected everything that happened in their world, but that knowledge of the planets and their influences could help them do everything from predicting the weather to curing diseases. Some even saw the planets as a source of magical power for manipulating their world, their fortunes, and the people around them. The planets were understood not as worlds in their own right, but as parts of one interconnected world with Earth at its center (or bottom), and God at the outermost sphere (or top). Far from being the best place in the world, Earth was a place of confusion and even torment—a prison of death and corruption as far from God as could be. The heavens, on the other hand, were perfect and unchanging—filled with power, and enacting divine will. Knowledge of the heavens was prized. As we will see, reuniting the soul with heavenly perfection through education became a goal for some medieval philosophers, and it guided the imaginings of those who wrote descriptions of the earliest voyages to the planets. It is in this light that we will come to understand Mars as the medieval mind knew it—through a discussion of astronomy, astrology, medicine, and magic, and finally through the writings of the medieval cosmic travelers Bernardus Silvestris and Dante Alighieri.

## MARS IN THE TIME OF THE BLACK DEATH

Consider an even more deadly plague than the one that faced us in 2020: the fourteenth-century Black Death. In October 1348, the Paris medical faculty delivered its report on the great wave of pestilence that was then sweeping over Asia and Europe, and which would ultimately kill as much as half of Europe's population. In its report, the medical faculty identified those things in the heavens and on Earth that had likely caused the plague. The plague, none of the esteemed men of medicine doubted, had been sent by the Christian God as a universal and indiscriminate punishment for the inherent sinfulness of humankind. But this was only the beginning of the explanation: the why but not the *how* of the plague. In the medieval Christian mind, though God was capable of intervening in whatever way he chose, his primary mode of acting upon the world was through the natural mechanisms of the world system he had created.

The heavenly cause of the plague was, they reasoned, "the universal and distant cause." It had been enacted through the combined influence of Saturn, Jupiter, and Mars. Here the doctors pointed to the date of March 20, 1345, when "there was a major conjunction of three planets in Aquarius." This event, along with previous conjunctions and eclipses, had caused "a deadly corruption of the air around us," signifying "mortality and famine." The medical men pointed out that since the time of Aristotle, conjunctions of Saturn and Jupiter had been known to cause "mortality of races and the depopulation of kingdoms." The conjunction of 1345 had been made even more cataclysmic by the involvement of a third planet: Mars. The effects of the conjunction were further amplified because Mars had been in the sign of Leo for several months in 1347.[1]

What exactly had Mars done? The effects of Jupiter and Mars in originating and perpetuating the plague had resulted from their qualities and the constellations through which they had traveled. Although the planets and stars were considered to be made of different stuff from the substances found on Earth—an aethereal fifth element unlike the earth, water, air, and fire that comprised the sublunar world—they nonetheless possessed qualities similar to those of objects found on Earth. All things in the heavens, like all things on Earth, were a combination of either hot or cold with dry or wet. Jupiter was hot and wet, and it met Mars in Aquarius, itself a hot and wet sign. This coincidence had amplified the qualities of Jupiter, enabling it to draw "evil vapours" up from Earth. Mars, also hot but excessively dry, had been able to "ignite" those vapors, "and as a result there were lightnings, sparks, noxious vapours and fires throughout the air."

Mars—described by the medical men as "a malevolent planet, breeding anger and wars"—had also contributed to spreading the poisonous vapors. In October 1347 it had begun moving in retrograde into the constellation of Leo, a fire sign that was correspondingly hot and dry. This had made the winter of 1347 unseasonably warm, and had caused strong winds to blow. These seasonal abnormalities not only drew more vapors up from Earth, but also helped to move them around the known world. These "alien vapours" had mixed with the air and corrupted it. When inhaled, the bad air carried corruption and rot into the body, destroying the "life force."[2] The medical community thus explained the plague as a divine punishment enacted by the regular and preordained movement of the heavens, influenced by the unchanging qualities of celestial objects, and enacted on Earth through unusual but still natural phenomena. Medical practitioners from the time of Hippocrates had understood that weather was directly related to the health of individuals and populations; airs and winds of different qualities could affect the body in different ways, and some weather could turn diseases into epidemics, amplifying the effects of effluvia produced in the bowels of the Earth by planetary influences.[3]

The will of God interacted with the human body through a well-ordered and intricately interconnected cosmos. Paraphrasing ancient Greek and Roman authorities, the astrologer Geoffrey de Maux explained in his own treatise on the plague that God had created the heavens and stars first, and had "endowed them with the power to rule all earthly matters, and because of this it can be said that everything which befalls us happens at the will of God; for it is God himself who moves the heavens and whatever is within them, and it is through this motion that there come all the chances of generation and corruption, and all the other chances which lie outside our free will."[4]

In medieval Europe, cosmology was not an independent discipline of knowledge but a central part of all knowledge, and central to the curriculum that formed the basis of a then new invention: university education.[5] The medieval understanding of disease was very far from what our modern germ theory would allow, involving aspects of astronomy, astrology, and meteorology—all of which were regarded as overlapping, if not intricately linked, areas of knowledge. Knowledge of the planets and their relationship to the world was central to the study of every natural phenomenon in the world and in the body.

### UNDERSTANDING MARS AND THE MEDIEVAL WORLDVIEW

Although the physicians appealed to the planets in explaining how the plague originated and spread, it is obvious that they ascribed neither will nor personality to the planets themselves. Mars may have been "malevolent," and may have contributed to the pestilence, but this was not because it had decided to do so. The influences of the planets resulted from their qualities, positions, and movements through the zodiac, as well as their connections to other parts of an intricate *machina mundi* (world machine) that structured the medieval European worldview. The fact that all of creation was connected—all part of one world with Earth and human beings at its center—meant that the movements of Mars and the other planets influenced every aspect of the natural world.

Just as with the ancients, understanding the medieval worldview means forgetting much of what we take for granted about our universe—that it is governed by gravity, mass, and energy, described by mathematical laws that appear to be the same everywhere, and defined primarily by a small amount of matter dispersed across an otherwise vast emptiness. We must instead adopt a model in which everything is everywhere connected, different objects obey different rules depending on their properties and position, and no empty spaces exist. To a medieval natural philosopher, the universe *was* the world—there was no other—and it was divided into a series of nested spheres (figure 4).

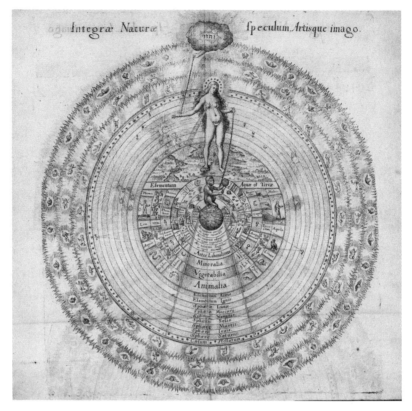

**FIGURE 4** The spheres of Earth and the heavens, as illustrated in Robert Fludd's *Utriusque cosmi maioris scilicet et minoris metaphysica, physica atque technica historia.* Courtesy of Science History Institute.

There were four primary elements: the element of earth belonged at the center of the universe, followed by water, air, and then fire. Where the sphere of fire met the sphere of the Moon, this marked the boundary of the sensory world, in which change, generation, and corruption defined existence. From the Moon outward lay the unchanging and perfect spheres of the planets and stars, where circles defined the natural motions of the heavens. Aether, the fifth element, composed everything from the Moon to the outermost edge of the heavens (even the seemingly empty spaces).

As the fourteenth-century Parisian scholar Themon Judaeus wrote,

"Every natural power of this inferior sensible world is governed by the heavens."[6] The planets were the cause of the movement of the four sublunar elements; their influence could change one element into another, combine elements into forms, bring things into being, and cause corruption to break things apart. Life was defined by change: it was birth, growth, maturation, degeneration, death, and decay. Without planetary influence, the elements would separate, eliminating all complex forms. Imagining a world in which planetary motion stopped was tantamount to imagining a world in which all elements would be frozen in place and no life could exist. Islamic, Jewish, and Christian natural philosophers saw this as the purpose for which God had created the heavens. According to the Islamic philosopher Averroes, "Heaven exists because of its motion; and if celestial motion were destroyed, the motion of all inferior beings would be destroyed and so also would the world."[7] The planets were an essential part of one divinely ordered world, not physical worlds in their own right.

Why were the planets connected to change in the physical world? This tradition, as we saw in the previous chapter, began in the ancient world. The ancient Greeks had counted seven planets in the heavens; ordered outward from Earth, these were the Moon, Mercury, Venus, the Sun, Mars, Jupiter, and Saturn. The Greeks had observed, as had other cultures before them, that the Sun had a very noticeable influence on the sublunar world. It supplied heat, brought the seasons through its movement through the ecliptic, and also was suspected to supply all of the light in the heavens. The Moon likewise had noticeable influences: it governed the tides, and its movements, phases, and appearance marked the times for planting, harvest, religious ceremonies, and other important annual occurrences. The appearance of different plants at different times of the year, the seasonal migrations of animals, and weather events could be associated with the movements or appearance of the Sun and Moon. It followed that just as the Sun and Moon were different in qualities and in influence, so too should be the other five planets.[8] The motion, light, and influences of the planets were understood to cause a variety of changes on Earth; they were believed to be responsible for invisible phenomena like magnetism, the spontaneous generation of maggots and other small creatures associated with decay, and the generation of different metals within the Earth. The material of the sublunar world was transmutable and held all potential forms, and these forms emerged and transformed because of the planets.[9]

For the philosophers, the planetary influences were products of the qualities of the planets. Each of the planets, like each of the sublunar elements, was a combination of hot or cold, dry or wet. The elements found in the sublunar world could be changed from one to another by changing one of their qualities. Water is cold and wet, but if we change one quality—if it becomes hot and wet—we produce air. Likewise, if it becomes cold and dry, we produce earth. If we take this cold and dry earth and instead make it hot, it becomes fire. These examples are overly simplistic; the actual transmutation of the elements was understood to be complicated, as alchemical recipes from this era testify, and the world was filled with countless forms that each required different balances of the elements. Nonetheless, we can understand how a planet like Mars, for example, exerting the influence of its hot and dry qualities upon the sublunar world, could, in conjunction with the Sun and other planets, participate in maintaining constant change.

There was at least one catch that natural philosophers had to overcome. The planets were made of aether, and aether was unchanging and could not contain within itself any mutable qualities. Strictly speaking, then, logic dictated that the planets could not have any *real* qualities. And yet, without these qualities, how would they influence the world? The answer to this dilemma was that the planets must possess these qualities *virtually*. The Sun, for example, was believed not actually to be hot. It had the ability to cause hotness only in the sublunar region. In this way it was "calefactive," able to heat only objects capable of being heated.[10] The Sun could not, it was assumed, heat the other planets, since they did not have the capacity to be heated. It could only illuminate them. Nonetheless, the Sun was considered the most noble of the planets and was often described as the king of the heavens, or as the heart of a living animal. Medieval natural philosophers agreed that the creator had wisely placed it in the middle of the heavens, below the superior planets Mars, Jupiter and Saturn and above the inferior planets Venus, Mercury, and the Moon. In this position, the Sun could govern and illuminate all of the heavens.

Rarely did the medieval natural philosopher question why the planets moved; their purpose within the divine world machine was virtually indisputable. But philosophers did argue among themselves about how and when the planets imparted their influences. Some speculated that the Sun gathered the planets' influences and transmitted them to the Moon so that they could then be relayed to the sublunar world; hence the significance of eclipses, when planetary influence was increased. Others believed that the Moon's influence prepared sublunar matter to receive other planetary influences. Still others speculated that the Sun and Moon together controlled the motions of the other planets—dictating, for example, when Mars's motion would become retrograde, and therefore where and how its influence would be felt. Some argued that the superior planets, like Mars, affected the permanence and duration of things in the world while the inferior planets affected the motions and corruption of those same things. In short, there was much to be debated about this perfect and eternal system.

### THE POWERS OF MARS

Medieval natural philosophers and physicians generally agreed that planets were "fortunate" or "unfortunate," "beneficent" or "maleficent." Mars was considered unfortunate and maleficent; in the *Ymago Mundi*, Pierre d'Ailly's popular fourteenth-century compendium of cosmographic knowledge, Mars is described as hot and dry, fiery and radiant, and therefore harmful and provocative of war. In his influential four-book treatise on astrology, the *Tetrabiblos*, Claudius Ptolemy had laid out the properties and powers of Mars in detail. He had explained first that the nature of Mars was "chiefly to dry and burn, in conformity with his fiery colour and by reason of his nearness to the Sun, for the Sun's sphere lies just below him."[11] Because of its dryness, Mars was male (along with the Sun, Saturn, and Jupiter; Mercury was both male and female).[12] On its own, the malevolence of Mars could bring about all kinds of ill effects. Its dryness could cause political problems such as war, tyranny, civil unrest, arson, murder, piracy, and enslavement. To the body it could cause symptoms as mild as fevers, or as severe as swift and violent death of the young. To the "condition of the air" it could bring hot weather, pestilence, lightning, hurricanes, and drought. It also could bring other problems such as the deaths of animals, plagues of locusts, and fires.[13]

Mars's influence could change when aligned with other planets, or when combined with the constellations of the zodiac. Like everything else, the signs had hot or cold, dry or wet qualities, and could either enhance or diminish the influence of a planet while it traveled through the heavens. According to the fourteenth-century French natural philosopher Jean Buridan, "A hot planet seems of greater power if it is in a hot sign than if it were in a cold sign because the sign and the planet can simultaneously influence heating, and thus a great hotness arises here below."[14] Hence the Paris physicians' focus on Mars in the sign of Leo, as a hot and dry sign, as a contributing cause of the Black Death. That Mars could produce exhalations from Earth to corrupt the air was related to Mars's normal ability to produce change within Earth, as its influence was understood to penetrate underground and produce iron and other forms.

Having studied the works of ancient Greek authorities like Hippocrates and Galen, in addition to more recent works by such Islamic thinkers as Ibn Sina (often latinized as Avicenna), medieval physicians knew that just as the sublunar world was made of four elements, the health of the body depended on the balance of four humours: blood, phlegm, and black and yellow bile. Just as the planets influenced the state of the four elements, so too did they influence the balance of the humours. Mars, along with the Sun, was associated with yellow bile and could cause diseases related to imbalances of this humour.[15] Each planet also was associated with the faculties—sight, touch, taste, smell, speech, thought—as well as the components of the body—bones, sinews, flesh, muscles, and so on.

The seven planets each had a special influence on the parts of the body. Scholars of ancient and medieval medicine call this form of astral or astrological medicine melothesia. Medieval physicians who practiced this art studied charts of the *homo signorum* (zodiac man) not only to help diagnose a patient, but to know what remedy to give and when (plate 2).[16] In his writings, Galen had set out a system the physician could use, in consultation with the astrological signs, to diagnose an imbalance in the humours, as well as to determine the time and location on the body for administration of the treatment. Ptolemy described the practice of reading the stars in the service of healing a patient, including the planets' associations with and effects on the human body. Mars was most likely to affect the kidneys, veins, gallbladder, genitals, and left ear. He declared, "Mars causes men to split blood, makes them melancholy, weakens their lungs, and causes itch or scurvy."[17] When aligned with Mercury, Mars could produce sore eyes and abscesses; the alignment could also rile the black bile and produce insanity. Aligned with Saturn, Mars brought disease and, under the right circumstances, pestilence. Saturn meanwhile, if not aligned with Mars, could help to fight the effects of Mars or another malevolent planet by increasing the efficacy of drugs and even the skill of the physician in determining what treatments to administer; this meant that there were good and bad times to seek medical help, as determined by the heavens. Mars rarely brought a good prognosis.

### TRANSLATION, TRANSMISSION, ASTROLOGY, AND MAGIC

Where did this worldview come from and why was this knowledge considered powerful? To the late-medieval natural philosophers, the signs of progress brought by their relatively new knowledge system were obvious. Let's try for a moment to see things from their perspective.

Europeans of the early Middle Ages recognized that they were far from being the most advanced civilization, and that they instead lived on the periphery of a larger, more culturally and technologically sophisticated world dominated by their more developed Eastern neighbors in the Middle East, China, and northern India—an idea that persisted well into the thirteenth century.[18] They believed this for good reason. The fall of the Western Roman Empire in the fourth century had isolated most of northern Europe, leaving it politically unstable, militarily vulnerable, and cut off from the eastern parts of the former empire. Education in the West, when it was available, was limited mainly to those who could retreat to Christian monasteries.[19] What remained of Greek (or "pagan") learning in early medieval Europe was scant, limited to fragments compiled in Hellenistic and Roman encyclopedias and handbooks, along with a small selection of Latin translations of Greek texts including an incomplete translation of Plato's *Timaeus* and parts of Aristotle's works on logic. These incomplete works became the basis for European learned wisdom and the introduction of the seven liberal arts in the early Middle Ages.[20]

Meanwhile, to the east, a very different story unfolded. The Islamic world emerged as an expanding and urbanizing empire that reached from Spain to India, including north Africa. Its cosmopolitan centers were places of learning.[21] Historians suggest that this moment saw an unprecedented "cultural explosion" that featured the translation, assimilation, and naturalization of ancient science, philosophy, and approaches to knowledge.[22] Libraries were built, embassies were sent in search of manuscripts, and scholars—including not only Muslims but Persians, Jews, and Christians—were employed to translate these texts into Arabic and make sense of them.[23] Active engagement with these texts and with the knowledge systems of their neighbors led to incredible advancements in many areas of Islamic thought—including the four interrelated areas of astronomy, cosmology, astrology, and medicine.

In eighth-century Baghdad, this new era of astronomy began with the translation of Sanskrit works into Arabic.[24] Greek and Greco-Roman astronomical writings—especially those of Ptolemy—would come to eclipse Indian ideas in the next century, but these Sanskrit texts nonetheless left their mark on how the Greek texts were read, as did Indian mathematical tools such as trigonometry, Indian methods for calculating parallax, and the adoption of Hindu-Arabic numerals.[25] Astronomers in the Islamic world adopted a multicultural approach, assembling elements from Athens, Alexandria, Constantinople, Syria, Iran, and India.[26] They worked in new institutions such as the *Bayt al-Hikma* (House of Wisdom), a library and academy in Baghdad intended as the center of the translation project and of the intellectual activity that went with it.[27] They began their own programs of observational astronomy and natural philosophy. Over the next two centuries, Islamic astronomers and skilled instrument makers made their own improvements to the African astrolabe—an analog computer that eliminated the need for trigonometric operations—and developed the equatorium, a device that could be used to calculate and predict planetary positions by turning dials, without the use of tables.

By the tenth century, Islamic scholars like al-Battani and Ibn Sina had improved on Ptolemy's *Almagest* by adding new observations made in Baghdad and Damascus (the accuracy of which would not be exceeded until the sixteenth century),[28] and appending new mathematical tools. They also made important new theoretical insights in planetary theory in their studies of the discrepancies found in Ptolemy's *Almagest* and his *Planetary Hypotheses*. The Abbasid court astrologer Abu Ma'shar wrote an *Introduction to the Science of the Stars* and its more influential abbreviated version, in which he provided an Islamic defense of the value of astrology and outlined its principles—a work that synthesized Greek, Indian, and Persian knowledge.[29] Islamic scholars like Ibn Sina took the first steps toward bringing Ptolemy's geocentric worldview and his astrology in line with the new monotheism, treating the planetary influences as a natural means through which God controlled occurrences on Earth.[30] Astrology, treated in a newly religious context, became a central concern of many Islamic works on astronomy.[31]

At the end of the tenth century, northwestern Europe began to reconnect with the world via trade. The eleventh century brought an age of holy Crusades, as the newly unified Christians of northern and southern Europe attempted to reclaim the Holy Land from the Muslims. One of the hopes they brought with them was that they would be able to convert the peoples they contacted in the Crusades, and in this way strengthen Christian culture through the incorporation of superior knowledge. The Crusades ultimately failed, but nonetheless reunited Europe with a larger world.[32] Perhaps counterintuitively, they brought Europe into more normal trade relationships with its Muslim neighbors, as it became part of the larger economic system still centered to its east.

The Crusades gave Europeans access to the storehouses of knowledge built by their Eastern neighbors. The next few hundred years were an exciting time for learning as European scholars assembled and made sense of a now expanded library of ancient texts. Aristotle's physics and Ptolemy's works on astronomy and astrology became the core of the scientific curriculum in the new universities that sprang up to accommodate a growing administrative state. This was not Greek knowledge per se, but a "Greco-Islamic storehouse of natural philosophy and science."[33] As we might expect from the story of transmission detailed above, Europeans reading these texts were heavily influenced by the commentaries and additions made by the Islamic scholars who had already spent centuries reconciling the texts with a monotheistic worldview, not to mention also making their own observations of the world around them.

These texts didn't just explain the world; they also suggested powerful ways of manipulating it. Medieval Europeans regarded the Eastern libraries they now drew upon—especially in centers of learning like Toledo—as centers for the occult, places where sorcerers studied magic and astrology. To be sure, there were magical texts transferred from the Islamic world to Christian Europe. Among these were texts of astral magic that featured spells or rituals incorporating the planetary spheres, calling upon the angels and demons of those spheres, and manipulating the unseen connections and sympathies that bound the parts of the world. This was done in the service of seeing into the future, seeking good fortune, achieving victory in battle, channeling astrological power toward the manipulation and influence of others, and a variety of other purposes. One of the most famous of these texts was the *Ghayat al-Hakim* (latinized as the *Picatrix*), a massive compendium of astrology and astral magic translated from Arabic into Latin in the thirteenth century. The *Picatrix* included prayers to the planets and stars. Some historians argue that in fact these magical texts, along with later alchemical works, held the first descriptions of experiments in manipulating the world via physical intervention.[34] In this way they can hardly be discounted as being outside the scientific tradition.

Later texts, such as the fifteenth-century *Liber de Angelis, annulis characteribus et Imaginibus planetarum* (Book of angels, rings, characters, and images of the planets), collected and perpetuated such experiments and bound them with collections of medical remedies. Medicine and magic both sought to effect changes in the natural world through the manipulation of natural objects and ingredients, so it makes sense that they might be found together. The experiments in the *Liber de Angelis* generally require one to make a ring of a specified material to represent a given planet or planets, to inscribe the name of the angels and demons of the planet or planets on the ring, and then, typically, to sacrifice an animal. The book describes what each planet can accomplish. Mars is "made for war and battles and destruction."[35]

For Mars, the experiments usually begin when one makes a ring of bronze to represent the red planet. The remaining steps vary. If you "sacrifice a bird of prey to the fire in the house, and write with its blood on its skin the name of the angel and the character," then "when you wish for armed soldiers to appear, or castles or mock swordplay, or that you conquer in battle, write the character and name of the angel on the ground and open the paper and it will appear and do as you wish." One can also sacrifice "a wolf or cat in a deserted place, write the character and name of the angel on the forehead," and thus have victory in any contest. Or, by sacrificing a rooster, writing the character and name of the angel on its skin, and then folding a penny in the paper, one can create a penny that will always return.[36] Many of these experiments end with burial of the astrological image the experiment produces, vesting its power in the Earth.[37]

Other variations follow, some of which include summoning or conjuring the demons from the spheres. To destroy anything one wishes, for example, one can create from bronze or red wax the image of Mars, then recite a conjuration of the "Red Fighter" (king of the demons of Mars) and his entourage.[38] Texts like *Liber de Angelis* make it clear that late-medieval Europeans combined the earlier forms of astral magic received from the Middle East, whose contents drew upon a variety of ancient source material, and contributed their own understanding of angels and demons, and so established a medieval tradition of necromancy.[39]

While Europeans feared occult knowledge and questioned whether it came from divine or diabolic sources, they nonetheless viewed astronomy, astrology, and magic as powerful and promising forces for understanding and intervening in the world. These arts could tempt the untrained mind to go too far in the pursuit of occult power, but they could also be put to good use when practiced within a proper Christian framework. Members of the clergy—who constituted the largest part of the small but growing fraction of Europeans who were literate and learned during this time—were the main practitioners of the types of astrology and magic found in these compendiums. Few doubted the efficacy of astrology or necromancy; they instead worried that those who lacked sufficient faith would be tempted or deceived by the very real demons they might contact through these experiments.

As hundreds of newly translated texts circulated across Europe, the occult influences of the planets were combined with the Christian worldview. As described above, the planets became understood as the natural means through which God enacted his will upon Earth; the lines between cosmology and religion were blurred. One example of a fourteenth-century rite used to remove worms from a vegetable garden combines the Holy Trinity with planetary influence, demanding that the worms leave the garden "through the Father, Son and Holy Spirit . . . through the Sun and the Moon and the other planets, through the stars and the sydera of heaven."[40] This prayer seems rather tame compared to the contents of *Liber de Angelis* or the *Picatrix*, with the absence of any names of demons, and no animal sacrifice other than that of the worms. Astrology, when properly used, was understood to be a perfectly acceptable means of understanding and even manipulating the world.

## SILVESTRIS'S JOURNEY TO MARS: A CREATION
## ALLEGORY TO OVERCOME THE FIRST DEATH

We have established that medieval Europeans regarded the planets not as worlds but as part of one world system. The heavens were sublime and perfect, made of one substance, an incorruptible aether shaped into a series of nested celestial orbs that carried the fixed stars and the planets in their regular and eternal circular motions. This ethereal region was separate in substance from the elemental region that existed below the Moon. If one took Aristotle's physics seriously—which most late-medieval natural philosophers did—then nothing made from the elements could travel into the heavens. If one did manage somehow to do so, one would find that the differences in the heavens—between spheres, planets, stars, and so on—were not due to differences in composition. The heavens were all aether, which had no sensory qualities or contraries, and no weight. Perhaps, many speculated, the visible planets were regions in the spheres in which the aether was more concentrated and became luminous through receiving or reflecting light from the Sun. Others believed some parts of the aether were more opaque for some reason other than density, but having to do perhaps with the virtual qualities bestowed by the creator.[41] But they were certainly not worlds upon which one could set foot.

So it might be surprising, knowing what we do, that medieval thinkers could imagine voyaging to the planets or witnessing them up close. And yet they did. They did so in ways that might not make sense to us now, and they saw things that we would not accept as accurate. And they did so in ways that reflect not only the shape and structure of the medieval worldview but the medieval view of the nature and purpose of human life, the mortality of the human body, and the eternal nature of the soul. And, perhaps not surprisingly, they did so in ways that evoked the ancient texts they drew upon, as well as their enthusiasm for the powers held within them. We will look at two voyages to Mars in this chapter. The first is in the *Cosmographia* of Bernardus Silvestris, which provides an allegorical account of the creation of the universe and allows us to see the parts of that universe through the experiences of the goddesses the author invokes to create it. We will finish this chapter with a discussion of the most famous medieval celestial journey ever written—Dante Alighieri's *Paradiso*.

A twelfth-century French Neoplatonist philosopher at the cathedral school of Chartres, Bernardus Silvestris, and his scholarly circle were engaged in the study of the newly translated ancient texts. His *Cosmographia*, which alternates between sections of poetry and prose, combined a preexisting allegorical tradition with new philosophical and natural knowledge. Within the text is an optimistic interpretation of the new knowledge and the role natural science can play in the corrupted human world.[42] We will spend some time getting to know the *Cosmographia* over the next few pages. But first we need to understand a bit more about the allegorical tradition with which Silvestris's work communicated, and its attitude toward the purpose of natural knowledge. In the process, we will get one brief bonus trip to Mars.

Silvestris's allegorical form was borrowed from the fifth-century Roman north African author Martianus Capella, who lived in what today is Algeria. Capella wrote the allegorical encyclopedia *De nuptiis Philologiae et Mercurii* (On the marriage of philology and Mercury)—a series of nine short books that would become incredibly influential to intellectual life in the early Middle Ages. In this work, Capella introduces the seven liberal arts in a personified celebration of the pursuit of knowledge through scholarship. The format of the first two books is, by Capella's own description, "mythical" or "fable," intended to prepare the reader's mind for the "true intellectual nourishment" of the following books.[43] In a story said to have been related to Capella by a muse, the maiden Philologia (love of learning) marries Mercury (intellect), and among the gifts given to the bride at her wedding are seven maids, each embodying one of the seven liberal arts—Grammar, Dialectic, Rhetoric, Geometry, Arithmetic, Astronomy, and Harmony. In the first two books, Capella describes the betrothal and wedding feast, followed by one book for each of the arts that compiles and attempts to explain the ideas of earlier authors.

The eighth book of *De nuptiis* describes what Capella knew of astronomy—which, interestingly enough, put the inferior planets Mercury and Venus not below the Sun but in orbit around it. But Mars makes its first appearance during the allegorical marriage feast, when Capella's muse first introduces astronomy. This example shows us allegory in action, and gives us a sense of its purpose. Here, responding to an offering of spices, the goddess Juno appears to the bride. Philologia requests that Juno "grant my request to know what goes on in the vastness of the sky."[44] In response, Juno lifts the maiden into "the citadels of the sky," where she shows her the spheres and the gods and godlike spirits who reside there.

The goddess and the maiden witness the Moon, Mercury, and Venus on their way to the Sun. After offering a devout prayer to the Sun, which appears as a ship from which "a fountain of celestial light poured forth, spreading in mystical emanations into lights which illuminate the whole world," the maiden is brought to Mars. Here she reaches what Capella identifies as the "circle Pyrois," thus associating the flaming horse of the Greek sun god Helios with the red planet. Phylologia must step over the mythical river of fire, Pyriphlegethon, to continue on her way to Jupiter; the river seems to flow from Mars down through the spheres to the underworld.[45] The influence of Mars is presented as a geographical relationship of shared natural and supernatural features, inhabited by natural and supernatural beings. Capella has used allegory to see what is otherwise unseeable—even as what is seen is not intended to be taken entirely literally—and has primed the reader's mind to receive more serious elements of cosmology. This approach will guide Silvestris in his own imagined journey to Mars.

Silvestris's other main influence was the fifth-century Roman scholar Boethius's De consolation philosophiae (The Consolation of Philosophy). As in Capella's allegorical tale, Boethius's story is based upon a fictional visit from a muse. De consolation was written while Boethius was imprisoned for treason and awaiting execution. In despair, he is visited by the muse Philosophia. The muse offers Boethius a correct understanding of the structure of the cosmos and the place of the human soul within it, claiming that this will cure his dejected spirit. She explains that her method is based on the revelation from Plato that all learning is recollection.[46] The soul already knows the cosmos, but has forgotten. To learn about the macrocosmos via astronomy and reason—to recollect the relationship of the human soul to the universe—allows the imprisoned soul to achieve harmony with the world.

How does the soul know the cosmos? Boethius's imprisonment is described as both real and metaphorical, a reflection of the Christian belief that mortal life is itself a form of imprisonment for the soul. Christian thinkers in the early Middle Ages were influenced by Plato's notion of the immortal soul, and by Greek philosophies and stories that described the movement of the soul through the stages of existence. Neoplatonists understood the split between macrocosm and microcosm—universe and body—to be the product of the descent of the soul from the heavens into the body. According to Macrobius, humans are condemned to suffer two deaths. The first, more terrible death is that of the soul when it "is being thrust from the radiance of its immortal state into the shades of death," and entombed in a body.[47] The body "represents the dregs of what is divine."[48] The influx of these dregs into the soul causes it to become intoxicated, and to forget all it has seen of the structure of the heavens during its descent. In fact, as Macrobius describes, the soul has seen and understood everything in its descent through the spheres, as the planets have bestowed attributes upon it during its journey: Saturn imparts reason and understanding, Jupiter agency, Mars boldness, the Sun vision and imagination, Venus passion, Mercury speech and articulation, and the Moon growth and fruitfulness.[49] It is only upon entering the sensory world that the soul, now fully forged, forgets these preparations.

Philosophia promises Boethius that through the study of the world, she can restore his memory of his soul's journey through the spheres and provide him with the harmony he seeks. She tells him that the pursuit of natural knowledge will fasten wings to his mind so that he can soar above the disorderly world into the heavens and recollect the order of the cosmos for himself. This restorative experience is captured in the poem "The Soul's Flight," which includes the verse: "If the road which you have forgotten, but now search for, brings you here, you will cry out: 'This I remember, this is my own country, here I was born and here I shall hold my place.'"[50] This notion that the soul is shown the heavens—its true home—before being entombed in a human body, expressed by Macrobius and Boethius, becomes an important element in Silvestris's allegorical journey through the spheres. Equally important is the hope that the study of the universe has the power to restore the human soul to its original dignity.

Silvestris's *Cosmographia* was designed to simulate a journey through the heavens and thus stimulate the reader's soul to remember and be restored.[51] Over the course of two books it recounts the exploits of four goddesses: Natura, Urania, and Physis, and their mother Noys, who is herself the daughter of God. Together the four goddesses create the universe and the first human. The first book, which Silvestris titled "Megacosmus" (Macrocosmos), begins as Natura complains to her mother that the matter of the universe is chaotic and unformed; she asks her mother to bring form and order to chaos. Agreeing that it is time to give form to the universe, Noys creates the four elements from primordial matter and imparts "the vivifying gift of soul" into the universe, creating the animals and plants. She separates the aether into the stars and heavens, creates the celestial sphere, and sets circular paths in motion.[52] We then come to the seven planets, including a description of Mars and its malevolent influence: "Mars, following next after the Sun, visits war upon proud cities, and his red glare often works strangely upon kings."[53] All of this is presented as an intelligent system of connection and influence: "The firmament learns from the divine mind, the stars from the firmament and the universe from the stars. . . . For the universe is a continuum, a chain in which nothing is out of order or broken off."[54]

In the second book, "Microcosmus" (Microcosmos), Silvestris focuses on the creation of the human. For this, Noys sends Natura to find her sister goddesses Urania and Physis. Urania is presented as the "starry queen," with expert knowledge of astronomy and astrology. Physis, on the other hand, has knowledge of medicine: she knows the heavens in order to understand their influences upon the plants and animals of the Earth, and she also knows how to diagnose disease and create remedies from plants and other natural ingredients.[55] It is Physis who creates the human, and she does so using the corrupted leftovers of creation. But Urania must descend through the heavens with the human soul, and in so doing she announces, "The human soul must be guided by me through all the realms of heaven, that it may have knowledge: of the laws of the fates, and inexorable destiny, and the shiftings of unstable fortune."[56] As in Boethius's narrative, Urania's description of the heavens offers recollection and restoration of grace. But she also promises something new, something neither Boethius nor Capella included in their allegories: the promise of astrology.

Urania and Natura, having met at the top of the heavens, must now make their own descent through the spheres to get to the work at hand. They pass through each of the heavenly spheres, including the sphere of Mars. Through their experience, the medieval reader was able to witness Mars. At the sphere of Mars, the two hear the sounds of a rushing waterfall, and see the "seething and sulphurous waters [of] the river Fiery Phlegethon" which issues forth from Mars. As in Capella's narrative, the fiery mythical river connects Mars to the sublunar world. Urania and Natura also see that Mars is "emboldened" as it passes through Scorpio, seeming to attempt to break from its orbit to appear "blood-red and terrifying" as a comet. The sphere of Mars "seemed so ill-governed and teemed with hostile vapors" that the muses decide to move on quickly to the Sun's sphere.[57] Silvestris's Mars is unruly, ready to break the order of the heavens and wreak havoc.

Silvestris's authorial approach to the Cosmographia is influenced by his predecessors, but his understanding of the disciplines he presents through his allegory is heavily influenced by the works recently introduced in the twelfth century. His medical knowledge, for example, shows influence from the eleventh-century physician Constantine the African, and his knowledge of astronomy and astrology show the influence of the eighth-century Persian astrologer Abu Ma'shar.[58] Twelfth-century scholars were excited about new astrological knowledge introduced from the Middle East, a fact reflected in Silvestris's astrological understanding of the power of astronomical knowledge.[59] With this knowledge comes a new sense of agency. Philosophia tells Boethius that he can harmonize his soul with the order of the cosmos and forget the ill fortune that has befallen him in the chaos of the sublunar world. Urania and Physis, by contrast, offer humans the dual weapons of astrology and medicine to combat the malevolence of the world.

Unarmed, and stripped of all knowledge of the cosmos, human existence appears almost as an afterthought to the rest of creation. But with this new knowledge, humans become the consummation of the creation, and full participants in it. Astrology and medicine were forms of agency in medieval cosmology.[60]

### MARS AND THE SECOND DEATH: ASCENDING THE SPHERES WITH DANTE

The most famous medieval journey through the universe is without a doubt Dante Alighieri's fourteenth-century *Divina Comedia* (*Divine Comedy*). A few significant differences separate Dante's journey from those addressed above. First, Dante's journey is not delivered to him by a muse. Instead, the *Comedia* is presented as a first-person account of a journey through the three realms of the dead. Second, while Capella and Silvestris both described the process of recollecting the downward journey of the soul to the body, and drew upon the creation narrative and notions of the soul they had found in Plato's *Timaeus*, Dante is more concerned with the second death described by Macrobius: the death of the body and the release of the soul. Dante's three-book journey begins in the sublunar world, and the poet is guided first by the Roman poet Virgil and then by the maiden Beatrice through the spheres of Hell, Purgatory, and Paradise. Third, thanks to the maturation of the translation movement in Europe, Dante is able to draw upon the Christian synthesis of Aristotle's physics and Ptolemy's model of the universe. The Christian Trinity, angels, and blessed souls occupy Dante's heavens.

But there are also similarities between Dante and his predecessors. For one, Dante was keenly interested in using his narrative to present the areas of knowledge that had formed his liberal arts education. Born in the late thirteenth century and probably educated in a monastery school in Florence, he had even more knowledge to draw upon in assembling his epic poem. Indeed, Dante's journey could only have been imagined by someone with a well-rounded education who held literature and history in at least equal regard to physical science; he was familiar with classical histories and literary works to which neither Capella nor Silvestris had access. Capella and Silvestris were in conversation with nature; Dante is in conversation with nature and history. The first two books of the *Comedia* are filled with notable Greek and Roman figures, and they combine Christian and Greek mythologies about the afterlife and the structure of the world. Historians have also argued that the *Comedia* was influenced by newly translated Islamic texts, including the *Kitab al Miraj*, which included ancient accounts of the Muslim prophet Muhammad's journey through the seven paradises to the edge of heaven.[61]

We won't spend much time here discussing the *Inferno* or *Purgatory*. We will simply note that one interesting aspect of the first two books, from a cosmological standpoint, is that Dante chose to structure the *Inferno* and *Purgatory* in a way that mirrored the spherical ordering of the heavens; each realm is composed of nine spheres plus one additional region. The historical figures we find suffering in the various spheres of hell, as well as their punishments, convey a Christian understanding of how sin and vice, along with divine justice, can be mapped onto the image of the well-ordered cosmos inherited from antiquity. Dante's vision of hell is both intellectual and terrifying; what worse fate could there be for one's immortal soul but to be released from its mortal prison only to be entombed within the world, prevented from returning to its celestial home, forever to be punished for succumbing to the various forms of human frailty?

Our focus here is primarily on Dante's journey into the heavens, and, given that Mars is the subject of this book, we will spend most of our time discussing what he sees when he encounters it. Dante and his guide, Beatrice, are permitted to travel through the sphere of fire at the edge of the sensory world, and through each of the nine heavenly spheres and the Empyrean. As this journey progresses, Dante learns lessons about the heavens similar to those learned by our previous authors. As he passes into the Moon's sphere in canto 2, for example, and can now see the nested spheres, Beatrice explains to Dante how the heavens work as a mechanism for doing God's will in the world. Dante also gets to see the outer edge of the heavens as he enters the ninth sphere of the Prime Mobile in canto 27. The movement of this sphere, which rests just outside the sphere of the fixed stars and is moved directly by God, causes all the spheres nested below it to move. And Dante even is allowed to enter the Empyrean in canto 33—the space outside the ninth sphere that only God can occupy. Feeling God's love for the universe and mankind, and sensing that his own will is "turned like a wheel, all at one speed," Dante understands why the heavens move. All of these observations summarize and express the Christianized view of Aristotle's physics and Ptolemy's spherical model of the universe.

But what of the planetary spheres themselves, and what of Mars? While the spheres of the underworld were each devoted to the classification of human sin, Dante learns that the spheres of the planets are each devoted to one of seven virtues. Mars is the planet of fortitude. Here Dante faces the challenge of describing how Mars—so long associated with war and malevolence—could represent virtuous fortitude. He dedicates Mars not to war itself, but to those who have died as warriors for God. He imagines the sphere of Mars to be occupied by those who have died as martyrs, who have given their lives as Christ did in the name of their Christian faith. This makes Mars one of the more significant planets in Dante's journey, occupying the subject matter of five cantos.

Dante's imagined experience of Mars was informed by what he knew of the planet. In a separate text, *Il convito* (*The Banquet*), Dante wrote extensively about the position of Mars in the heavens. Here he departed somewhat from other thinkers who placed the Sun at the center of the heavens, as the fourth sphere among seven. For Dante, Mars was at the center as the fifth sphere among nine, including those of the fixed stars and the Prime Mobile. Here he wrote that, because of its central position combined with its hot and dry qualities, Mars could appear as a ball of fire, flaming red, with bright vapors.[62] So it's not surprising that, as Dante approaches Mars in the *Paradiso*, he is overwhelmed by its brightness and must look away. Beatrice helps him to regain his composure, and he can look at Mars and see that "the smiling star" is "red as fire," and "beyond the customary red of Mars."

Coming closer to Mars, Dante sees the planet resolve into many sparks or rays of light that make a cross—a symbol of the ultimate martyr, Jesus Christ (plate 3). The cross is animated by the souls of the martyrs, which "sparkled radiant" as they move along the cross. The experience is also musical, as Dante hears an unknown hymn of holy devotion, which speaks of rising and conquering enemies of the faith. From amid the lights that make up the cross, one of Dante's martyred ancestors comes to speak to him, explaining the significance of the cross to Mars, and offering to name the martyrs who make up its light. As the famous martyrs are named, their lights flash like lightning and move on the cross. Dante describes visual experiences including "a splendor thrust along the cross" and a "flame wheel round itself" as he watches attentively, like "a falconer who tracks his falcon's flight."

It is also while Dante is in the sphere of Mars that we are reminded, through the mention of Dante's impending exile, of the purpose for this journey. Dante will return from the realms of the dead to bring to the human world an accurate account of the structure and purpose of the cosmos. It is significant, then, that he chose to write and publish the *Comedia* not in Latin, but in vernacular Tuscan Italian. That the *Comedia* was accessible to literate Italian readers not educated in Latin, still a small segment of the total population in the fourteenth century, is no doubt one reason why it is still today one of the most famous and influential texts of medieval European literature. As we will see in the next chapter, it influenced imaginary journeys into the heavens for centuries to come—even after early telescopic observations began to reorder and transform the planets from ethereal points of light into worlds.

# *RESTRUCTURING THE WORLD*

For Europeans at the end of the fifteenth century, the world grew not only in size but also in diversity of all kinds, including that of human cultures. For the Indigenous peoples the European explorers encountered in the Americas, this was the prelude to centuries of colonial rule and erasure. The riches of the Americas became the wealth of Europe, at great cost to native peoples. And it wasn't just gold and silver that enriched Europe; it was also the new experiences and information about the world Europeans brought back with them. The classical view that Europe brought civilization to the world is backward in many respects. Historians today criticize this Eurocentric view, arguing, for example, that "it was not that Italy found the world and then transformed it; rather that the more advanced Eastern world found Italy and enabled its rise and development."[1] We've already seen how medieval European economic and intellectual progress was due largely to multicultural exchange with Africa, the Middle East, India, and China, and a connection to the larger world economy these partners dominated. This trend continued as Europe turned toward the Americas. The culture that would define the Renaissance, the Enlightenment, and the Scientific Revolution developed as Europe stood "on the threshold of an expanding world of multicultural contact."[2] Moreover, Europe was not unique in its experience of this dramatic increase in knowledge and scientific practice; many other parts of the world participated in and contributed to this transformative moment.[3] The new astronomy emerged alongside a new Eurocentrism that reframed encounters and exchanges as "discoveries," and it began to draw new maps of heaven and Earth with new centers and peripheries.

The new age of encounters was bound to bring intellectual change. For the next three centuries, the world machine of late medieval natural philosophy began to break down. Alternatives to the Aristotelean-Ptolemaic system that dominated the previous centuries emerged and competed as a growing community of astronomers and natural philosophers debated the true nature and physics of the cosmos. Mars and the other planets became worlds—places the imagination could not only visit but inhabit. And Earth slowly broke loose from its mooring, joining the planets in orbit around the Sun. But this didn't happen quickly. Even when the red planet did change, it didn't immediately lose the qualities and influences that ancient and medieval people had attributed to it. Early modern scientists at the end of this period could talk seriously about Mars as a physical place because shared conceptions about Earth and its relationship to the heavens were destabilized, not erased.

Astronomy played a key role in chipping away at the old worldview. But some of the most important changes that took place in the European understanding of the heavens had to do with the center of their cosmos: Earth. We are going to begin at the center and see how ideas about that world informed the voyages across the Atlantic Ocean that would begin to decenter it. And we will look to the sky to see how earlier observations of comets and novas had already opened up a space for multiple models of the heavens to proliferate. We will visit briefly with Galileo Galilei and the astronomical community before taking a trip to Mars with the Jesuit scholar Athanasius Kircher to see how he made sense of the new discoveries in the heavens.

## CHANGES AT THE CENTER: MARS AND THE GEOGRAPHY OF THE INHABITED WORLD

Let's begin with a globe. This globe is known as the *Erdapfel* (German for "Earth apple"), or as the Behaim Globe, to those who wish to give credit to its originator, Martin Behaim. It is the oldest known globe in existence (figure 5).[4] Although it is ostensibly just a globe of Earth with some glaring errors and omissions that reflect the limited geographic knowledge and tools of the late fifteenth century—it was, after all, one of the last globes produced before the discovery of the Americas—it is in fact a globe of a different world that no longer exists today. This is true because of the information and ideas it embodies, not because of the details it lacks.

The *Erdapfel* represents the beginning of the end of the old understanding of Earth and its relationship to the universe. This is the pre-planetary Earth, the center of a cosmos in which Mars's primary purpose was to work in concert with the rest of the heavens to shape and change the sublunar world. New discoveries on Earth and in the heavens over the next two hundred years would chip away at the logic and textual authorities that made up this worldview. But at the time when this globe was made, this image of the world seemed very stable.

For more than three hundred years before the *Erdapfel* was produced, European scholars and theologians discussed and debated their world, generating not a complete consensus but a robust set of questions and answers that attempted to explain the world in which they lived. This long conversation took place in texts that were painstakingly written out and copied by hand in the monasteries and universities of the late Middle Ages. The explosion of printed books beginning in the mid-fifteenth century brought these texts and their commentaries to an even wider audience of what we might anachronistically call urban middle-class readers.

Historians today estimate that by 1500 CE the commercial presses that sprang up to meet the new market for books had produced as many as a million of them—probably a greater number of books than had been produced in Europe since the medieval period began.[5] The majority of the European population was still illiterate, but books were now abundant and affordable enough that they could be found outside palaces, monasteries, and universities. After 1500, printed books, along with even more affordable woodcuts, pamphlets, and leaflets, would increasingly be used to proliferate new images and ideas. But in these

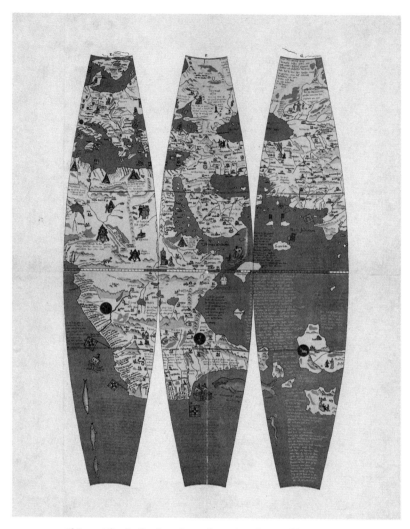

**FIGURE 5** This 1908 facsimile of sections of Martin Behaim's fifteenth-century *Erdapfel* shows not only the world but the people who inhabited it and what was known about them. University of Wisconsin–Milwaukee Libraries.

initial decades of printing in Europe—perhaps for as long as the first century of printing—the majority of texts produced were faithful reproductions of the late medieval European worldview.[6] Mars was a part of this worldview; it was a piece of the divine machinery that affected the shape and character of Earth and its inhabitants.

The knowledge of the world depicted on maps and globes was based on a variety of sources. But three of the primary sources came from Africa, written by the Greco-Roman scholar Ptolemy, whom we've already met. Ptolemy contributed to the medieval worldmakers three influential works: a treatise on astrology in his *Tetrabiblos*, a mathematical and predictive model of the heavens and planetary motion in his *Almagest*, and an atlas with instructions for mapping Earth's surface in his *Geographia*. While the *Almagest* and *Geographia* are agreed to have been influential in the making of maps and globes,[7] it was the *Tetrabiblos* that explained who and what existed where on the globe, and why.

The *Almagest* described the nature of Earth and how it could be divided into four quadrants. The inhabited world—which occupied parts of the three known continents of Europe, Africa, and Asia—takes up only one of the four quadrants. In book 2 of the *Tetrabiblos*, Ptolemy considers an astrological ethnography of the various parts of the inhabited world. Ptolemy carved the inhabited world up into climatic zones, based primarily on their distance from the tropics and the North Pole. The characteristics of the plants, animals, and humans of these regions were determined by their climate, which was in turn primarily determined by the influence of that most important planet, the Sun. For those living in the Mediterranean and at similar latitudes above and below the equator, including Ptolemy's home in Alexandria, the conditions—neither too hot nor too cold, and never experiencing violent shifts in temperature—produced humans ideally suited for astronomy and philosophy. They were "in general more shrewd and inventive," "better versed in the knowledge of things divine," "sagacious, investigative, and fitted for pursuing the sciences specifically called mathematical," all because "their zenith is close to the zodiac and to the planets revolving about it."[8]

The people living in this temperate climate were more fortunate than those who lived in the colder climate up north, who by comparison were pale, tall, and well-nourished, but savage. Those living in the southern reaches of the inhabited world, meanwhile, were inclined to be short, dark-skinned, sanguine, and savage. These descriptions of "national characteristics" were no doubt based upon prejudices toward the people with whom the Greeks and Romans had contact, or about whom they had heard stories, and the influences attributed to the Sun and planets were added after the fact.

Ptolemy also addressed the influence of the stars and planets on the different quadrants of the inhabited Earth in his *Almagest*. Different quadrants would have different relationships to the zodiac and the planets that moved through it, and these differences would manifest in similar ways to how "some peoples are more inclined to horsemanship because theirs is a plain country, or to seamanship because they live close to the sea." "Special traits" would dominate different regions, depending on the dominant signs and planets that most influenced that area.[9] Ptolemy divided the twelve signs of the zodiac into four triangular arrangements, each containing three astrological signs, and went on to address the effects of each on the regions of the inhabited world. Again, he assigned the most favorable traits to those who lived in the central portion of the inhabited Earth—which he no doubt regarded as the center of civilization.[10]

Mars played an important role in defining the differences between human populations. For those lucky enough to live at the center, the subtle influence of Mars gave them qualities of leadership. For the northern parts of Europe, including Britain, Gaul, and Germany, the region's closer familiarity with the sign of Aries and the planet Mars made the people "fiercer, more headstrong, and bestial." In the East, a similar familiarity with Aries and Mars made the people "bold, godless, and scheming." Meanwhile, in parts of the northeast quadrant, Mars, because of its junction with Venus and the influence of the sign of Capricorn, made the women kind and affectionate to their husbands. Elsewhere, because of other influences, Mars produced a variety of traits both admirable and detestable—boldness, knavery, and treachery, to name just a few. Mars could produce magicians, impostors, deceivers, carnivores, warriors, and more, depending on the signs and other planets with which it was combined. Even after the world system that supported them had been picked apart, these ideas were used to justify the enslavement and exploitation of people around the world. They remained the basis for a Eurocentric cosmological explanation for the politics of a colonial world order.[11]

Then there were the riches. To early modern Europeans, place and its relationship to the world machine also determined what might be found there. Place was a principle of the universe. Place was "the medium through which celestial bodies imparted form to sublunary creatures."[12] The movements of the heavenly spheres transformed the elements and brought them into new combinations to become compound forms—including metals and precious stones. Geography, with its attention to place, participated in the philosophical endeavor not only to explain but to predict how and where these changes might take place.[13]

As we already know from the previous chapter, the Sun was the planet considered to be most responsible for changes on Earth. It was thought to have influences in addition to light and heat that could not be sensed by humans, but which could nonetheless penetrate Earth's surface and bring about changes in its interior. The other planets also had occult influence that could penetrate Earth; Mars was understood to be able to produce iron. Some medieval thinkers, like Albertus Magnus, had reasoned that because the planetary spheres were arranged with the Sun at their center (in the fourth sphere), this most noble planet might be the vehicle through which all planetary influences were collected and transmitted.[14] The Sun wasn't the center of the world, but it was still the most noble planet and the center of the heavens.

To the south of the inhabited world was what Ptolemy and the medieval geographers who followed him described as the Torrid Zone—a place between the tropics too hot to be inhabited by humans (and possibly also plants and animals) because it was too much under the influence of the Sun. For this same reason, however, it might be rich with precious metals and gems. The Europeans had noticed that their southern neighbors seemed exceptionally rich, and some had concluded that the Sun—the great cosmic engine of transformation—was responsible. The tropics must be a veritable garden of material wealth.

The *Erdapfel* is probably our best visual indication of where Christopher Columbus thought he was sailing when he headed south and then west across the Atlantic. Columbus and Behaim both evidently believed that the world's diameter was smaller and that Asia extended farther east than turned out to be the case. Where Columbus encountered the islands of the Bahamas, he and Behaim calculated he would reach the "Indies" of the South China Sea.[15] Here Columbus believed he would find the great wealth that ancient and medieval astrology told him the hot climate would produce. He set sail on August 3, 1492, carrying not only maps and charts but ideas gleaned from his own heavily annotated copies of Pierre d'Ailly's *Ymago mundi*, perhaps the most influential medieval compendium of cosmography and astrology, along with Ptolemy's *Geographia* and Marco Polo's account of his travels in Asia—all of which also informed Behaim's globe.[16]

But, as we know, Columbus ended up somewhere new and unknown to Europe. Earth was much larger than had been imagined, and its inhabitants and climates defied ancient and medieval understandings of planetary influence. Those Europeans who now traveled into the tropics themselves, such as the Jesuit Jose de Acosta, had their own very physical encounters with the expectations of his scholastic education. Everything Acosta had read prepared him for a violent and unbearable heat when he arrived at the equator. And yet it was so cold that Acosta had to sun himself to stay warm. He later mused, "What could I do then but laugh at Aristotle's *Meteorology* and his philosophy?"[17] This was just one of the ways in which experiences of the tropics and the New World strained the expectations of the natural philosophers.

By the time Nicolaus Copernicus's *De revolutionibus orbium coelestium* (*On the Revolutions of the Heavenly Spheres*) was published in 1543, the gears of the medieval world machine were already starting to slip—a little. The transformation from the Aristotelian-Ptolemaic world system to a new one was neither abrupt nor as revolutionary as has often been argued. It would be more than a century from the date Copernicus's heliocentric ideas appeared before they were widely accepted. Even Copernicus himself saw his heliocentric system as a revision of the old system, not an erasure. He offered no new physics, no new explanation of how the planets moved or why—only arguments, primarily from classical sources, for why the new arrangement made logical sense.[18] Furthermore, he likely made his heliocentric argument in defense of the credibility of astrology, as a way of giving the old tradition new astronomical foundations.[19] And Gerardus Mercator's more accurate terrestrial globes, though they rendered Ptolemy's *Geographia* obsolete, remained connected to celestial globes that denoted the enduring connection between Earth and the heavenly spheres.

But let's not get too far ahead of ourselves.

### CHANGES IN THE HEAVENS: NEW WORLDS IN THE SKY

What impact did the encounter with a New World on the other side of the Atlantic Ocean have on the later reconceptualization of Earth as just one world among a system of worlds orbiting the Sun? The answer to this question is complicated. The tensions between ancient authority and new experiences did set the stage for the emergence of the new astronomy. However, scholastic natural philosophy had already spent hundreds of years questioning the ancient texts, accommodating them to a world and a religious context their original authors had not anticipated, and had found them to be an incredibly malleable framework for incorporating new knowledge—even when they could be shown to be flawed. This, at least for a time, did not change.

Even if the Aristotelian-Ptolemaic worldview didn't immediately fall to pieces when Columbus brought back news of his voyages, the European approach to knowledge making was nonetheless transformed by this experience. The commercial expansion of the Spanish empire engendered new contexts for empirical practices that would come to dominate early modern science. Colonial authorities had to assess, test, and validate knowledge about new plants, animals, and commodities, and had to establish new networks and institutions within which this massive undertaking could be carried out. The economic and political necessities of this period of colonialism and commercial expansion legitimized empirical practices to an unprecedented degree. The new practices, which permeated a variety of areas including navigation, natural history, ethnography, and instrument making, proved exceptionally valuable in the mastery of new territories, peoples, and Indigenous knowledge systems.[20] This led to the emergence of a new empirical culture and, along with it, one of the key shifts in knowledge making that distinguishes the rise of early modern science: the shift from authoritative knowledge and logical argument meant to *explain* the world, to *useful* knowledge obtained through personal experience about how nature worked and how it could be exploited.[21] Because of the economic and political significance of this new enterprise, geographical discovery became the metaphorical language in which new observations and ideas were described throughout Europe well into the seventeenth century, and even to this day.

Changes observed in the heavens did contribute to the destabilizing of the old worldview, although not necessarily to geocentrism. The Danish astronomer Tycho Brahe—reputed to be the best naked-eye observer of his day, and certainly one of the best funded—observed the nova ("new star") of 1572 and the comet of 1577. Using some of the best instruments in Europe, Tycho determined that both phenomena occurred not in the atmosphere, but well beyond the sphere of the Moon. The appearance of these new celestial objects—and the disappearance of the nova after two years of observation—began conversations about change in the heavens that defied notions of perfect, crystalline spheres. Perhaps the heavens were fluid, Tycho suggested.

Tycho also reasoned that the heavens needed some rearranging. Like many sixteenth-century astronomers, he believed that the Ptolemaic system and its various revisions had become cumbersome. Making the planets orbit the Sun would bring them into better alignment with his own planetary observations. But while Tycho hailed Copernicus as a hero for his mathematical approach to planetary motion, he didn't like Copernicus's model for multiple reasons. For one, Tycho didn't believe that Earth, made as it was of heavy and solid elements, could be in motion like a planet. He also reasoned that the movement of Earth relative to the fixed stars as it orbited the Sun would generate observable stellar parallax. He had never detected parallax in all his years of observing and recording positional information for eight hundred stars in the night sky. A moving Earth with no observable parallax, by his calculations, would put the fixed stars at a distance from Earth seven hundred times greater than that of the most distant planet, Saturn.[22] This seemed like an unreasonable amount of empty space. Furthermore, based on his own measurements of the relative sizes of stars, Tycho argued that putting them at such a great distance would make them unreasonably massive. He proposed a compromise that combined elements of the geocentric and heliocentric models and preserved Aristotle's distinction between the heavens and the sublunar world: the planets orbited the Sun in a heaven that was fluid, not crystalline, while the Sun, Moon, and fixed stars orbited the central Earth. From an observational standpoint this was no different from Copernicus's model, but it avoided the problem of parallax.

Famously, in 1609 Galileo produced his first telescope. He had been tinkering with the Dutch instrument for more than a month, and had then spent the winter of that year observing the heavens. By December he had built a telescope that could magnify by a power of twenty. With this instrument, he looked at the Moon and found its surface to be rough, imperfect, and covered in remarkably earthly features like mountains and seas. At this power of magnification, he could only view about one quarter of the Moon at a time. Employing his knowledge of perspective, as well as his artistic skills with pencils and watercolors, he produced composite portraits of his observations of the Moon in its phases. Galileo's lunar portraits are almost caricatures in their outsized depictions of the shape and scale of the Moon's features.[23] But these illustrations were part of Galileo's rhetorical strategy in what became the *Sidereus nuncius*. In the *Nuncius*, Galileo showed his readers a new world:

one with mountain peaks taller than their terrestrial counterparts, and seas as large as Bohemia. He presented himself as the man whom God had graced with the honor of seeing the ancient but obscured lunar territory for the first time, and of giving him the talents to make it visible for others to see.

Galileo's *Nuncius* is written not to provide mere descriptions of the objects he discovered in the heavens, but to provide an illustrated narrative of the experience of his observations.[24] While the text is presented from his firsthand perspective as the viewer-protagonist, the narrative invites the reader to imagine themselves as the one with their eye to the telescope. His singular experience of discovery becomes a shared experience. In this way, Galileo's narrative approach employs tricks one can find in earlier accounts of the discovery and exploration of the Americas, and to similar vicarious effect. But there is a difference: Galileo's *Nuncius* presents a new mode of exploration, of seeing and experiencing remotely.[25] The reliance upon sense experience and instruments would become a hallmark of modern science, but that would come later. At least for now, for Galileo's audience in 1610, this narrative of discovery was connected to a history of exploration.

The very same Venetian presses that published Galileo's *Nuncius* had built their fortunes publishing accounts of the exploration of the Americas. These presses had created textual theaters of imagination through which Europeans could participate in the exploits of explorers and adventurers.[26] More Europeans experienced New World exploration through texts and maps printed in Venice than did so firsthand. The Age of Encounters came to dominate the local cultural and intellectual landscape for much of the sixteenth century preceding Galileo's telescopic discoveries. It's not surprising, then, that Galileo was often compared to Columbus in his moment of triumph. Italy had produced a cosmic Columbus, one who had been led by God "along paths never trod before by the human mind" to become "the discoverer of a new world of stars."[27]

This language of geographical discovery and of new worlds must have landed well with Galileo's new sponsors, the powerful Medici family in Florence. Florentines, having brought Ptolemy's *Geographia* to Europe in the fifteenth century, could claim to have set the course for Columbus's voyage of discovery.[28] When the Medici secured their power over Tuscany, Cosimo I incorporated the Americas into the family's identity as prosperous conquerors. Italian ideas of geographical discovery, virtual exploration, and the colonialization of knowledge became important parts of the family mythology. For the family's growing collection of American artifacts, Cosimo I commissioned the construction of a new collection and display space, the Guardaroba Nuova, at the Palazzo Vecchio.

The Guardaroba was a large trapezoidal room containing a cycle of fifty-three geographical maps of Earth on the front of a series of finely crafted walnut storage cabinets designed to hold the family's growing collection of artifacts from around the world. A Medici-centered cosmography of the known universe, the room contained maps, globes, painted constellations, illustrations of flora and fauna, and portraits of historical leaders. It held the largest armillary sphere in the world—a massive representation of the Aristotelian universe with Earth at its center—built under the supervision of the Pisan astronomer and professor of mathematics, Antonio Santucci.[29] The room also featured an impressively large terrestrial globe that can still be seen today in the Palazzo's Sala delle Carte Geografiche. The room linked Earth to heaven, knowledge to power, and the Medici family to exploration and discovery. Mixing maps and images with the material objects collected from the Americas and elsewhere, the room demonstrated what parts of the world were known to the Medici, as well as what was owned—if not physically, then intellectually—by the family.[30]

If one had an ambition of offering new discoveries for the Medici cosmography—as Galileo did, dedicating his discoveries to the family in pursuit of their patronage—the Guardaroba was perhaps the room to study. By the time Galileo forged his relationship with the Medici, they possessed one of the largest collections of New World objects and visual depictions of the Americas in Europe. While the family collected artifacts from other parts of the world as well, the artifacts from the Americas "provoked a new drive to catalogue, document, and represent the world at the Medici court," and to contribute to the production of new knowledge about the world.[31] The 1608 marriage of Cosimo II, to whom Galileo would dedicate his new discoveries, featured an allegorical enactment of Amerigo Vespucci's voyage to America. Not long after this enactment of the Medici's heritage of discovery, the family became patron to the newest Italian discoverer of worlds.

### THE NEW MARS

Galileo and his fellow Copernican Johannes Kepler both believed in the urgency of convincing others about the reality of the heliocentric worldview. However, neither of them made it past the Moon in their imaginings of what the other planets—if they now became worlds—would look like. Galileo observed changes in the apparent size and brightness of Mars's disk as it moved in its orbit. Kepler famously struggled with Mars's irregular motion for eight years, and from this struggle derived his insight that the planets moved in ellipses, not circles. But when Kepler took his dream journey beyond Earth in his posthumously published *Somnium* (1634), it was the lunar surface upon which he landed. Just because these champions of the Copernican worldview didn't imagine trips to Mars, this doesn't mean that no one did. We can still get to Mars by shifting our focus to the Jesuit scholar Athanasius Kircher.

Kircher arrived in Rome in 1634, in the wake of Galileo's trial and subsequent house arrest, to take up the position of chair of mathematics in the Jesuit Roman College. Here he inherited the instruments left behind by his predecessor, Christopher Clavius; and, as Clavius had, he continued to teach the Ptolemaic model of the universe. Perhaps because of the humiliation of Galileo, Kircher did not dare publish his own cosmological treatise during his first twenty years in Rome.[32] In fact, he denounced the theories of Copernicus and Kepler in his 1641 treatise on magnetism.[33]

While Galileo was connected to the colonial world and the idea of geographical exploration through the presses of Venice and the metaphor of discovery, Kircher was connected through a network of Jesuit missionaries stationed around the world. In Rome he constituted a central hub in this network, and from it he had access to objects and knowledge produced through multicultural encounter and exchange. His style of inquiry was wide-ranging and encyclopedic, but his two main interests were the natural world and Egyptian hieroglyphs and artifacts; he made no discernible distinction between these two areas of inquiry, and he worked on both projects simultaneously.[34] He saw the hand of God at work in experiment and observation. Learning about the world led to knowledge of God. This meant that all ancient cultures who had uncovered truths about nature had also known the Christian God, even if they had not yet been offered salvation. He assembled an eccentric selection of natural objects and cultural artifacts that served this argument into a small but ever-expanding museum at the Roman College.[35]

Kircher had a keen interest in astronomy, and his studies combined empiricism with religion and occult philosophy.[36] As a young priest stationed in Avignon before coming to Rome, he had turned a room in the Tour de la Motte of the Jesuit college into an astronomic clock within which he used mirrors to reflect sunlight and moonlight onto walls on which he had traced the constellations, signs of the zodiac, and hours of the day. The room also depicted the latitudes of towns in the tropics, and included meridians for determining the correct time at different locations. The images of Saint Francis Xavier and Saint Ignatius on the left and right of the meridians connected this world clock to the Jesuit mission abroad.[37]

Kircher also observed eclipses, studied the sky with telescopes, made his own observations of sunspots, and produced and published depictions of the Moon, Jupiter, and Saturn in his *Ars magna lucis et umbrae* (1646), a book dedicated to his ideas about light and optics. He described the Moon as "a Globe substantially like our Earth, a body, that is, of earth and water, that is, one composed of heavenly Water and Earth."[38] Kircher's Moon was subject to changes; he instructed his reader not only that it had features similar to Galileo's, but that they might see "a smoky exhalation" on the edge of the lunar disk.[39] Kircher celebrated the telescope and the new discoveries in the heavens—as well as the exploration of the sublunar world and the opportunity to bring Christianity to all of its peoples—as evidence that Christian Europe was transcending the natural knowledge of the ancients, and was therefore coming to a truer understanding of God.[40]

In 1657, Kircher published an imagined journey to the planets in his *Itinerarium exstaticum* (Ecstatic journey)—a story told in 288 pages of baroque Latin dialogue, followed by a 173-page treatise expanding on the same topics (figure 6). Like Kepler's *Somnium*, the journey takes place in a dream, and like Dante's *Paradiso*, the trip requires a divine escort. It was intended as a fictionalized treatment of Kircher's own ideas about the universe—not, he assured his reader, the report of a divine revelation.[41] It is in this work that we see Mars as a world for the first time— and we set foot on it.

Kircher's story begins at the conclusion of an evening concert at the Roman College. Our narrator, a Jesuit scholar named Theodidactus ("Taught by God"), remains in the hall and is lulled into sleep when the musicians retune their instruments to replicate ancient musical scales and resume playing. In his sleep, Theo meets a beautiful but frightening angel named Cosmiel who has shimmering wings of every color, hands and feet that "surpass jewels," and eyes like burning coals. In his right hand the angel holds a celestial globe depicting the planets, in the other a jeweled measuring stick.[42] Cosmiel folds Theo up in one set of wings and, with a second set, carries him into the heavens. Kircher's heavens are fluid, as Tycho suggested; Theo sees the liquid rays of sunlight traveling through the liquid aether.

It is Cosmiel—whose demeanor is blunt and sharp-witted—who delivers Kircher's ideas about the heavens to the questioning Theo, as Theo delivers a first-person narrative describing in vivid imagery the journey through the universe and the worlds he encounters.[43] One idea Kircher introduces through Cosmiel is the importance of experiment. Aristotle and the philosophers got a lot wrong, Cosmiel insists, because they did not experiment; ideas alone cannot give one a complete

FIGURE 6 Athanasius Kircher's protagonist and his angelic guide set out to tour the celestial spheres. Wellcome Collection.

understanding of the world.[44] The type of experiment Kircher favored was the analogical experiment. This was not the hypothetico-deductive experimental method we value today, but rather the small-scale replication of natural phenomena based on connections built into the fabric of the world. In an earlier work, for example, Kircher had suggested that one could simulate the effects Mars has on a person by placing them in a room with a burning globe made from resin, arsenic, sulfur, pitch, antimony, mercury, and gold pigment. This would induce in the subject mania, frenzy, fever, and inflammation.[45] The experiment speaks to ancient beliefs about the influences of the planets on humans, as well as seventeenth-century practices in alchemy that combined artisanal knowledge of recipes for bringing about changes in nature (not limited to the transmutation of metals) with ideas about matter, its properties, and its propensity for change.[46]

That one could simulate the effects of Mars using a recipe of earthly materials speaks, on the one hand, to the transformation of the planets from condensed spheres of aether to worlds made of stuff similar if not identical to what was found on Earth. This led many, including Kircher, to a Neoplatonic version of the universe in which the heavens were made not of an incorruptible fifth element, but of the same four mutable elements that defined the sublunar world. But it also spoke to the idea that the universe, no matter how it was arranged, was an everywhere connected whole. Kircher's most famous demonstration of these invisible connections was the magnet, which made visible the "hidden knots" that bound everything in the universe; these knots were products of God's divine love.[47] Kircher believed that these connections did not produce the whole, but rather were products of the totality of the creation. There was no point to looking for laws of nature, and no deeper reality to mathematical descriptions of natural phenomena, though they could be useful. Kircher was interested in the sympathies and influences that comprised the fabric of the universe.[48]

The Mars that Cosmiel and Theo visited on their voyage was not based on any known features of the planet. Writing in 1657, Kircher would only have had access to vague descriptions of what others had observed through their telescopes during oppositions. Beginning in the 1640s, multiple astronomers had reported seeing dark patches on the planet's surface, or patches with less reflectivity than the rest of the planet's disk. No detailed drawings had yet been published. Kircher's Mars was his own invention.

As Theo and Cosmiel travel from the Sun to Mars (the book does not discuss the arrangement of the heavens, but the planets are visited in the pre-Copernican order), Theo begins to feel the planet's malevolence. While Dante was overcome by the brightness of Mars, Theo is struck by "the incredible stench of its vapors." It appears as a "hellish globe," its surface "roiled by a huge whirling vortex." "Stop! Stop!" he pleads. "Don't take me any nearer! I can tell that globe holds evil things in store for me." Not only does Mars offend his senses, but Theo also feels the influence of Mars on his thoughts and emotions. "There's some kind of deep-seated fury inside me," he tells Cosmiel, "and I seem to be burning with a sense of rage and indignation."[49] These ill effects—which correspond to those attributed to Mars in Ptolemy's *Tetrabiblos*—are relieved when Cosmiel fills his eyes, ears, and mouth with a heavenly tincture that puts him at ease and allows him to breathe normally in the planet's foul air. Cosmiel explains that the planet is made of a "violent influence" and is "conflicted by a battle of the elements of war."[50] Indeed, as we soon find out, Mars is a pretty unpleasant place to visit.

Preexisting notions of the planet's qualities and astrological influence are now made physical. Much as in Kircher's earlier Mars experiment, Theo describes the planet as being made primarily of sulfur, bitumen, arsenic, and naphtha. Cosmiel sets Theo down on the rim of a volcano, from which he surveys the Martian landscape. The surface is rocky, jagged, mountainous, and covered in fireball-spewing volcanoes and lakes—surface features in keeping with Capella and Silvestris's celestial river of fire, but now global in scale and no longer allegorical. Cosmiel next takes Theo to the edge of a flaming ocean of pitch and bitumen, rough and scaly in appearance, and porridgelike in consistency. It is the most disgusting place Theo has ever been, and he remarks that "you'd be better off calling this a mountain of flaming pitch than a planet." Theo is once again gripped by the desire to leave Mars, and naively asks why such a planet would exist in nature: "Who could stand the foulness of this globe? Who is spewing out this smoke, this breath of pestilent vapors, this repulsive stench?"[51]

In words reminiscent of God's speech in the Book of Job, Cosmiel reminds Theo that Mars, like the more beautiful and benevolent planets, is a work of God and part of the universe's greater whole: "Don't call what God has set up with such foresight for the preservation and ornament of the universe names like 'disgusting' and 'pestilent'!" Like Job, Theo concedes. He allows Cosmiel to show him the entire globe of Mars. Here we learn that it is the dry and fiery exhalations of the planet—not the color of its surface, which is covered in black soot— that gives Mars its red appearance when viewed from Earth. Theo is also careful to explain that the smells and the substances he describes to the reader are not literally the same as those found on Earth, but that he has been "forced to use our familiar terms" for substances unknown to him. Before departing for Jupiter, Theo witnesses the apparition of a swarm of fiery knights on horseback, brandishing swords and whips of fire in their hands. Cosmiel reassures Theo that these are not demons, but servants of God—protectors of Mars and ministers of divine justice.[52]

Kircher's Mars seems strange to us today, and it's fair to say that it was probably strange in its own time. His Martian geology, if we can call it such, aligns with Galileo and Kepler in its treatment of the planet as a physical world with Earthlike, if hellish, surface features. But Kircher's view of Mars was also clearly informed by his understanding of astrology, alchemy, and other medieval knowledge systems built to explain the unseen tapestry of occult connections and affinities between every part of the universe. His journey takes place in a fluid cosmos that, following Tycho, acknowledges the mutability of the heavens. But its depictions of the substance of Mars and the angels who dwell there invoke the learned magic of the Renaissance. Its treatment of planetary influence seems to come almost directly from Ptolemy. The planets can be visited, but their significance is still tied to their connections to earthly events. And Kircher's understanding of why malevolent planets like Mars exist in the heavens to cause such distress draws upon Jewish and Christian understandings of the Bible.

But lest we think Kircher was entirely unique, we know that at least one of his Jesuit colleagues, Valentin Stansel, chronicled a similar journey to the planets from his colonial post in Brazil. In his account, rather than flying with an angel to the planets one after another, Stansel is carried by a muse in dreamlike voyages to one planet at a time, departing and returning each time to a lush tropical garden. Stansel's Mars has no surface features; it was made from aethereal matter. Nonetheless, Stansel gets a biblical explanation for why Mars is allowed to wreak such havoc on Earth. His guide, Urania, reminds him that the serpent was present in the Garden of Eden. Is there not still divine justice in the universe, despite the evil that exists there? And hasn't God, in his divine wisdom, also provided the other more benign planets to temper Mars's influence? And doesn't the hot and dry influence of Mars also temper the cold of Saturn or the humidity of the Moon?[53]

### OVERLAPPING WORLDS

The versions of Mars that Kircher and Stansel experienced were each unique to them. In important ways, however, Kircher and Stansel's versions of the red planet demonstrate something more general to this period: in many minds, the cosmos was now in flux. These versions of Mars were products of two uneasily coexisting views of the universe. On the one hand, Mars was still part of a living and intelligent world machine in which every part of the universe had a role to play in a divine plan of maintaining and controlling life on Earth. On the other hand, new ideas that would later be refined and pieced together into the materialist view of the universe found insecure footing. The pieces that comprise the heliocentric solar system we know today didn't begin to comprise a coherent whole until the end of the seventeenth century.[54] The observations of Galileo and other astronomers might hint that the planets were worlds like Earth, but no philosophy explained the physics of why or how other worlds existed. Why would they move in circles or ellipses, if they weren't made of a special fifth element or held in their orbits by crystalline or fluid spheres? Why would Earth move if circular motion was not natural to the elements comprising it?

From our standpoint, Galileo's work is a foundational moment in the development of modern science. But to many of his contemporaries, his immediate influence on the direction of science was not so obvious. Natural philosophy valued explanation. Galileo had offered no explanation for *why* the planets, including Earth, should orbit the Sun. His correspondence demonstrates that he did try to come up with an explanation as early as 1615. It was based upon his observations of sunspots, and drew upon astrological notions of the Sun's occult influence, for which he found justification in biblical passages. The movement of the sunspots demonstrated that the Sun rotated; perhaps they were also evidence that the Sun was consuming a power bestowed upon it by the fixed stars. Perhaps this power then emanated from the Sun in a "spirited, tenuous, and fast substance" and penetrated everything in the universe, causing the celestial objects to rotate around the center. And perhaps this entire system was what was spoken of in Psalm 19, which, Galileo noted, states that "God hath set his tabernacle in the sun."[55] Perhaps the hero astronomer wasn't so different from Kircher after all.

Kepler claimed the laws of planetary motion from his victorious struggle to define the orbit of Mars. Finding the elliptical path was so difficult it that at times he felt like he was struggling with demons.[56] But even he had no recourse to anything other than Aristotelian philosophy as he attempted to replace the shattered physics of the spheres with terrestrial physics of motion in a fluid heaven. He explained Mars's motion by imagining it was pushed by imperceptible spokes with the Sun as their fulcrum, and that the size and shape of the ellipse, as well as the speed at which Mars traveled, was a result of a polarized magnetic relationship between the Sun and Mars. But Kepler had no explanation of what kept Mars from simply leaving its orbit and traveling away in a straight line. Having invoked the Platonic solids in his explanation of why the six planets, including Earth, would be arranged in such a way around the Sun in the Copernican worldview he championed, he fell back on a commitment to the sacred geometry of the heavens, even as he questioned so many other such commitments.

As the seventeenth century progressed, the worldview that Kepler's mathematics glimpsed would come more fully into being, even as the Aristotelian logic and Neoplatonic geometry Kepler employed in his work fell out of favor. A new natural philosophy aimed to strip matter and its motions of any sense of soul or intelligence. Characterizing this transition, historians describe the decline of Kircher's type of "scientist-magician, whose knowledge is of the occult powers of nature,"[57] and for whom nature was divine and inscrutable, as being eclipsed by mathematical physicists like Isaac Newton, who insisted that "nature operates according to mechanical principles, the regularity of which can be expressed in the form of natural laws, ideally formulated in mathematical terms."[58] The universe was thus transformed from a unified living system into a clocklike machine. Describing natural phenomena, as well as classifying and ordering nature's parts, became more important than explaining the divine logic that bound together the parts of the universe.

And while it seems as though we've now moved far from the story of geographical "discovery" with which we began this chapter, we have in fact come full circle. This turn to measurement and classification was accelerated, if not directly stimulated, by the new demands of colonial imperialism.[59] One of the recognized founders of modern science, Francis Bacon, made his famous proposal for a new empirical study of Earth and the heavens at the same time as England began its own expansion into the "New World." For his own part, Bacon did not shy away from the metaphor of geographic discovery: "Surely it would be disgraceful," he wrote in his 1620 *Novum organum*, "if, while the regions of the material globe—that is, of the earth, of the sea, and of the stars—have been in our times laid widely open and revealed, the intellectual globe should remain shut up within the narrow limit of old discoveries."[60] As we will see in the next chapter, geographical discovery remained the metaphorical language in which new observations and ideas were described throughout Europe, and later in America. This has direct implications for Mars, as the first attempts to map the red planet's surface coincided with the push to map the interiors of those parts of the world Europeans considered remote. Knowledge about Mars remained tied to knowledge about Earth, as well as to the political and cultural forces that defined centers and peripheries.

# THE MAKING OF MODERN MARS

A new Mars—a "modern" Mars—connected to Earth not through influence but through a shared origin and evolutionary path, would emerge by the end of the nineteenth century. Beginning in the Enlightenment, the red planet transformed under the influence of new philosophies of nature. New physical explanations for planetary motion made Mars subject to the same laws and forces as Earth. New theories that explained Earth's history as a planet in the cosmos put the two worlds in the same family. New telescopic and spectroscopic observations of Mars, combined with theories of planetary, biological, and cultural evolution, brought a new Mars into focus. Many believed this Mars was an Earthlike planet and likely home to its own forms of life, perhaps even intelligent life—and some sought evidence for the existence of that life. With these developments, vast and varied fictional versions of Mars proliferated.

In Europe and in America, scientists and popular writers celebrated the idea of progress. The evolution of the Sun and the planets became just the first part of a long history culminating in the superior civilizations of Western Europe. Discussions of Mars were swept up in this history. European and American astronomers applied the same geographic tools and methods to Mars as explorers did to the parts of the world they considered remote. Similarly, in imagining what peoples might inhabit Mars, and how intelligence would be expressed, these Martian geographers applied the same prejudices they did to the peoples of Earth. They sought not only signs of life, but signs of well-ordered hierarchical societies. While some writers of Mars fiction during this era echoed these prejudices, others, like H. G. Wells, questioned them by imagining how Europeans might be treated by a superior race of Martians seeking their resources. These fictional accounts—along with cautionary tales about debates over Martian life—set the tone for robotic exploration of Mars in the ensuing space age.

### MARS GETS ENLIGHTENED

We will spend most of this chapter in the nineteenth century. But first, a brief rundown of the developments that made Mars a physical world is in order. The French philosopher René Descartes (1596–1650) provided one of the first fully mechanical descriptions of the universe that managed to describe how, without aethereal physics or crystalline spheres to hold them in their orbits, the planets could be made to orbit the Sun. Drawing upon a physical phenomenon familiar to anyone who has watched water flow through a gutter or down a drain, Descartes built his world system around the rotating vortex. His universe was a plenum, everywhere filled with matter in motion. Motion in the plenum was defined by an interlocking set of vortices, centered around every star in the universe. God had filled the plenum, created the vortices, and initiated the universe's perpetual motion.[1]

Earth was no longer special—at least not in its composition. Mars now became the equal of Earth, made from the same types of matter, broken into infinitely divisible particles, and obeying the same laws of nature. Nor was our solar system special. It was located in one of many vortices. Descartes's model made it possible to imagine other planetary systems orbiting the other stars—each with a vortex of its own and containing the same matter. Now equating planets with worlds, philosophers in the seventeenth century commonly speculated about life on other planets—either in our solar system or in others.

Bernard Le Bovier de Fontenelle (1657–1757) in France and Christiaan Huygens (1629–95) in the Netherlands both suggested that the Cartesian vortices implied innumerable such worlds and gave us no reason to doubt their inhabitation.[2] Huygens was convinced that a benevolent and rational God would not create such a universe filled with planets without also creating intelligent beings to make use of each and every world. He did not wonder whether there was life on other worlds, but instead speculated about what that life might be like.[3] Mars attracted Huygens's attention; and in 1659 he drew a rough sketch of the planet, based on his own telescopic observations, that included an hourglass-shaped sea—the first recognizable surface feature on Mars. This was surpassed in detail by drawings the astronomer Giovanni Cassini (1625–1712) published in 1664. Cassini also managed to track the movement of the hourglass sea, and used it to determine the length of the Martian day to within a few minutes, noting that it was only forty minutes longer than the length of a day on Earth. Mars was starting to become more Earthlike.

The English physicist, mathematician, and alchemist Isaac Newton (1642–1726) provided the mechanical philosophy that ultimately came to define modern physics. To Descartes's matter and motion, Newton proposed adding a third category: natural forces.[4] The mathematical description of the universal force of gravity in his 1687 *Philosophiæ naturalis principia mathematica* (Mathematical principles of natural philosophy) was Newton's landmark achievement, and he imagined that other similar forces, also lawlike and universal, were responsible for such phenomena as magnetism and electricity.[5] Newton's forces acted at a distance, without direct contact, and this roused the skepticism of other mechanists who saw it as a return to the occult forces of Renaissance natural philosophy. These criticisms weren't far from the mark. Like that of Descartes, Newton's ultimate cause was God; but Newton's God did more than set things in motion. Matter in Newton's universe was essentially inert and passive, but that universe was God's "sensorium," in which he was present in all places at all times. God's laws could thus act universally without interruption across time and space.[6]

Newton was hardly a model mechanist. He devoted three decades of his life to alchemical studies, rebelling against the constraints of absolute mechanism.[7] While he believed in the mechanical and mathematical expression of nature, like Kircher he also believed that there were deeper truths hidden in the world by its creator. Going beyond mathematics, Newton attempted to explain what caused gravity, borrowing at least one concept from his alchemical practice. Gravity was caused by "active principles" bestowed on the material of the universe by God at the time of creation.[8] These principles operated in matter, and caused bodies to be attracted to one another in proportion to how much matter each body contained. Hence, they were measurable. As Newton wrote in proposition 7 of his *Principia*, "Gravity exists in all bodies universally and is proportional to the quantity of matter in each."[9]

Newton's universal law of gravity removed Earth from the center of the universe. It allowed planets like Mars to have their own centers and to obey the same physics as Earth, rather than a separate physics of heavenly bodies. The Sun could be the center of the solar system, with the planets orbiting in ellipses around it, and each planet could have its own gravity in relation to the amount of matter it contained. Gravitational attraction could shape each planet from chaotic "parcels" of matter into globes. But the planets had not formed through strictly natural processes. The laws that governed matter were God's will, and God played an active role in his universe. Newton didn't believe that the laws of nature alone were sufficient to create or maintain the solar system. He imagined that the Bible's book of Genesis fit somewhere within this work.[10] He suggested, "This most beautiful system of the sun, planets, and comets, could only proceed from the counsel and dominion of an intelligent and powerful Being."[11]

Into the eighteenth century, God still appeared in explanations of the universe. However, God's role as designer or clockmaker was gradually replaced by a view of the universe as an unfolding plan. God no longer established order from chaos, but had stepped aside long ago, having created matter and natural laws that could establish order on their own. Newton's successors attempted to discover new laws that governed forces and processes in other realms of nature.[12] New forms of calculus developed in Europe by mathematicians like Leonhard Euler translated Newton's physics into the new language of celestial mechanics, and mathematically inclined astronomers and theoreticians either let go of their objections to the idea of action at a distance, or modified Aristotle's invisible aether into a medium through which forces could travel (something Newton himself had done to explain how light traveled through space).[13]

## GEOLOGY, ASTRONOMY, AND THE NATURAL HISTORY OF WORLDS

The understanding of Mars during this period can hardly be separated from the new understanding of Earth. Enlightenment geologists searched for a "theory of the Earth"—the story of the natural creation of the third planet, as well as its relationship to the other planets of the solar system.[14] Earth's condition as a planet with a history as a member of a family of planets, as well as its susceptibility to universal natural forces, became important aspects of understanding its origins and development over time (figure 7).[15] The first such theory of the Earth appeared in the French natural philosopher Benoît de Maillet's (1656–1738) posthumously published *Telliamed* (1748).[16] A devoted Cartesian, Maillet suggested that the evolution of a planet involved a long process of drying out, as it slowly lost surface water to the vortex—a process that might take billions of years.[17]

Only a year after the publication of the *Telliamed*, Georges-Louis Leclerc, Comte de Buffon (1707–88) published his own theory of the Earth. Buffon proposed a scenario in which Earth and the planets had originated from a comet's collision with the Sun. This hypothetical event might have produced hot globules of solar matter that could then congeal into the planets.[18] Buffon did not claim that this was how the solar system had come into being. He insisted only that this scenario proved that a naturalistic story *could* account for the formation of a planetary system. He argued that the uniformities of the solar system—the common revolution around the Sun in the same direction and in nearly the same plane—pointed not to the work of a greater intelligence, but to

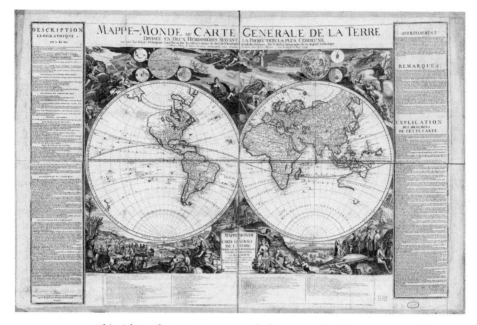

**FIGURE 7** This eighteenth-century mappemonde shows expanding European geographic knowledge of the interiors of the continents, and connects this knowledge to astronomical observations of the Sun, Moon, and planets. Library of Congress.

one common natural cause. What this cause was, he did not claim to know; but it need not be the hand of God.[19] Buffon's hypothetical story of solar system formation made all the planets the result of the same event, and all subject to the same directional process. Each planet had to cool from its initially molten state. After the crust solidified, great deluges created an ocean so deep that all of Earth's surface was under water, with the global ocean eventually retreating to reveal the continents, as well as the fossils of marine animals.[20]

Similar theories of natural creation took root in philosophy. Toward the end of the eighteenth century, Pierre-Simon, marquis de Laplace (1749–1827) began his speculative exploration of the history of the solar system in similar territory as that of Buffon. He noted the well-known fact that all of the planets and their satellites rotated in the same direction and nearly in the same plane. This, he insisted, was most likely the result of one common initial cause. The most likely explanation he could imagine was that all the planets had once been part of the same rotating mass. If the Sun had once had a hotter atmosphere extending as far as the orbit of Saturn, he speculated, this may have cooled over time, contracting into rings and then planets. The accretion of the ring material into planets was a hot process, which meant that each planet had begun its evolution as a hot molten sphere before entering a long process of secular cooling.[21]

Laplace's speculation seemed to find confirmation in the work of the German-born British astronomer William Herschel (1738–1822). Herschel, who became famous for the discovery of the planet Uranus in 1781, considered himself to be a "natural historian" of the universe. He was not content to simply observe and describe the objects of the heavens. His work on nebulae, which he conducted in collaboration with his sister Caroline Herschel (1750–1848), was focused on identifying and classifying a set of natural types that he could arrange in an ordered series reflecting the natural laws at work in the universe.[22] In this sense, Herschel was attempting to apply to celestial objects a set of practices used on Earth in the search for patterns in nature, the goal of which was to uncover the order in nature's plan. Over the course of their career, the Herschels discovered and classified thousands of new nebulae into four main types. By 1791 they were able to construct from these types what historian Simon Schaffer describes as a "temporal and dynamic natural history of the heavens" driven by the force of gravity.[23] In the sequence, a cloud of nebular fluid condensed to form stars, these stars would attract additional nebular material to form comets and planets, and the stars would eventually explode, returning nebular material to the universe to begin the sequence again.[24]

Herschel's general belief was that all planets were likely inhabited, so long as the conditions for life were favorable. In 1776 he examined the Moon through his telescope and reported seeing vegetation. During the Mars opposition of 1781, he turned his attention to the red planet, measured the length of the Martian day for himself, and performed an extensive survey of the polar regions. He also for the first time proposed a special relationship between Earth and Mars. Based on the similar length of their days, the tilt of their rotational axes, and their nearness to each other in the solar system, he speculated that Earth and Mars were perhaps most similar to each other of all the planets: the two planets likely had similar seasons driven by the melting and refreezing of ice at their poles. Herschel went on to suggest that Mars had an atmosphere similar to that of Earth, and that Mars's "inhabitants probably enjoy a situation in many respects similar to our own."[25]

### THE NINETEENTH-CENTURY COSMIC EPIC

By the mid-nineteenth century, modern science had set Earth and Mars on a common path. Whether one believed the solar system was created by God, by cosmic accident, or by the regular activity of matter, motion, and force, Earth and Mars were now at minimum members of the same family of planets and subject to the same natural laws. At most, they were nearly identical twins. Scientists and popular writers in the nineteenth-century British Empire, now the world's dominant naval and colonial power, would assemble this mixture of speculative suggestions into a new directional cosmic narrative that reflected a Victorian ideal of progress. This new narrative combined ideas of the physical evolution of the universe, solar system, and planets, with the development of life, intelligence, and Eurocentric notions of modern civilization. The historian Bernard Lightman has dubbed such progressive monad-to-man narratives as "evolutionary epics."[26]

Our first stop in our search for Mars in the nineteenth century is an author who was specifically interested not in Mars, but in the history of Earth and the rise of European civilization. Writing in the middle of the century, the Scottish publisher Robert Chambers (1802–71) is credited with giving the epic its popular form, by writing one of the most widely read books of his day. In his anonymously published *Vestiges of the Natural History of Creation* (1844), Chambers took on the persona of "a bourgeois Virgil in Dante's Divine Comedy" as he led his readers on a historical journey beginning with the birth of the Sun.[27]

Much of what Chambers presented was not original to him. His skill was in synthesizing, summarizing, and presenting the work of others in a coherent and diverting story. Chambers was fascinated, as were many of his readers, by the notion of progress. His own study of phrenology and associated race science had led him to search for the natural causes of everything, including human intellect, culture, and civilization—and this included what he perceived to be the natural dominance of European civilization.[28] Not surprisingly, he presented the "Caucasian type" as the highest form of human development, with the other races described as "representations of particular stages" in human development.[29] What was novel in Chambers's work was that his deftly crafted narrative wove ideas and theories from different disciplines (science had by now begun to professionalize and divvy up intellectual territory) into a natural history of progress that culminated in the ascendance of the British Empire as a world power. Bound up in this story was the history of the planets. Described in the first few chapters of *Vestiges*, their development is presented in terms Laplace or Herschel might have recognized. The planets were "bound up in one chain—interwoven in one web of mutual relation and harmonious agreement, subjected to one pervading influence which extends from the centre to the farthest limits of that great system, of which all of them, Earth included, must henceforth be regarded as members."[30]

Other popular writers carried this thread forward. Richard Proctor (1837–88) emerged as one of the most widely read astronomers and popularizers in the second half of the century. His first major success was a treatment of the question of life on other planets called *Other Worlds Than Ours* (1870), and it was followed by a series of works mainly related to planetary astronomy. Proctor's view of the solar system owed some debt to Chambers for popularizing the idea that the planets had all been produced and shaped by the same set of causes. However, where Chambers had removed God from all but the beginning of the cosmic evolutionary epic, Proctor made God present in the unfolding of the universal plan. Proctor's thinking on the question of life on other worlds, for example, was teleological. Like Huygens before him, he believed that the creator must have had some purpose in mind in creating a universe filled with stars and planets. For this reason, Proctor's brand of thinking can be labeled "cosmic, post-Darwinian natural theology." For Proctor and several other popularizers, the clues to God's ultimate plan were to be found through the instruments of modern science.[31]

One of Proctor's favorite examples in presenting his case for the likelihood of life elsewhere in the universe was Mars. An era of mapping Mars had begun as early as 1840, when the German astronomers Wilhelm Beer and Johann Heinrich von Mädler first applied nineteenth-century geographical techniques to depicting the planet's surface.[32] They weren't alone in attempting the decipher the surface of Mars; the French astronomer and popularizer Camille Flammarion estimated that by 1877 at least 391 drawings of Mars had been made since the invention of the telescope. In his own work, Proctor privileged the drawings of his fellow countryman William Rutter Dawes, and in 1867 he made his own geographical projections from those drawings. Proctor populated his map of Mars with the names of mainly British astronomers, giving it its first set of place names. He also employed findings from the new science of spectroscopy, first applied in astronomy in the 1860s. In 1867 Sir William Huggins attempted to probe planetary atmospheres, and concluded that Mars's atmosphere must contain water.[33]

Through the visual technology of the map and his compelling narrative style, Proctor presented Mars to his readers as a miniature of Earth. He confidently asserted that the dark spots on his map were seas and the lighter spots continents, that the atmosphere was like our own, and that Mars enjoyed much the same climate as Earth. Who could doubt, he asked, that the "Martial lands are nourished by refreshing rainfalls; and who can doubt that they are thus nourished for the same purposes as our own fields and forests—namely that vegetation of all sorts may grow abundantly."[34] Through the telescope, Mars looked like Earth. Astronomers like Herschel had already emphasized the many points of similarity between Earth and Mars. And the spectroscope provided all the evidence Proctor needed to conclude that the red planet was indeed very much like Earth. He drove the point home with an imaginative scene of a Mars populated by intelligent beings: "Mars presents a scene which requires no very lively exercise of the imagination to dot with villages, towns, and cities, peopled with busy workers."[35]

## CANALS AND CARTOGRAPHERS

The year 1877 marked a turning point in thinking about the surface features of Mars. It was in this year that the Italian astronomer Giovanni Schiaparelli (1835–1910) spent the better part of eight months observing Mars using an eight-inch refractor on the roof of Milan's Palazzo di Brera. Ignoring the names Proctor had given the Martian features, Schiaparelli introduced the nomenclature that would come to dominate Mars geography in a set of latinized names drawn from the classical and mythological geography of the ancient world: names like Hellas, Elysium, Amazonia, Cydonia, and Tharsis—names still used today. But it wasn't just the names that departed from previous maps of Mars. Schiaparelli saw new features; his Mars was covered not in shapeless subtle shadings, but in sharp-edged and intersecting lines dividing the landscape into islands. Schiaparelli suggested that these lines were *canali*, an Italian word that can mean either canal or channel. Schiaparelli was himself undecided on the question of whether Mars was inhabited, and whether these were features of intelligent design. But he noted optimistically that life was not out of the question: "As far as we may be permitted to argue from the observed facts, the climate of Mars must resemble that of a clear day upon a high mountain [on Earth]."[36]

Schiaparelli continued to observe Mars during subsequent oppositions. In 1894 he wrote that he had observed the polar caps of Mars, and suggested that they were made of snow and ice. He also claimed to see seas and continents. And he suggested that water was distributed by "a network of canals, perhaps constituting the principal mechanism (if not the only one) by which water (and with it organic life) may be diffused over the arid surface of the planet."[37] This was, he concluded, "a true hydrographic system."[38] He still did not go so far as to say that this system was artificial, but he was adamant that the possibility should be considered. Natural processes related to the evolution of the planet could explain such a system, but so could "extensive agricultural labor and irrigation upon a large scale."[39] The famous astronomer's writing was very suggestive. But many astronomers believed that the lines were cracks in the surface of Mars caused as the planet cooled and solidified during its evolution.[40]

None of the observers who mapped Mars during this period actually saw the planet in the way their maps presented it. The movement between observation and cartographic representation is mediated by the process of projection. This is especially true for observations of Mars, which through even the best Earth-based telescopes are fleeting and fragmentary. For lucky observers, Mars is resolved only in very brief flashes. These flashes are not of the entire surface of the planet, but only parts. Mars astronomy "engaged the observer in a strobic contest to collect data in the brief moments when both the telescope's apertures and the atmosphere through which it had to penetrate allowed optical clarity."[41] Observers had to "learn" to observe Mars, to wait patiently until they thought they had seen the same thing appear multiple times, and to then draw only what they now believed to be a real feature of the planet. Rendering Mars required "waiting intently for a moment of still air, then quickly recording an image before the memory could fade."[42] Each sketch typically captured only one or two features. Only once these had been compiled and projected onto the coordinates of a map or globe did a geographic depiction of Mars emerge.

Despite their shortcomings, topographical maps of Mars became valuable cartographic icons in staking and communicating scientific knowledge claims during this period (figure 8).[43] The process of mapping, and the use of a terrestrial vocabulary of seas, channels, islands, and continents, not to mention canals, reinforced the idea that Mars was Earthlike. While astronomers disagreed publicly in their interpretations of the features they saw on Mars, nearly all took for granted the idea that Mars was physically Earthlike, and most agreed that it was a living world. The maps they generated allowed them to speak with more certainty about the red planet's habitability than the observational or spectroscopic evidence alone. The maps also connected planetary astronomy to the imperial practice of exploration, the context in which

**FIGURE 8** Five globes of Mars made in the nineteenth and early twentieth centuries represent a range of representations of the planet's difficult-to-resolve surface, from a land of continents and seas to one of vast deserts and canal systems. Image © Whipple Museum of the History of Science, Cambridge (Wh.1268; Wh.6622; Wh.6625; Wh.6238; Wh.6211).

all nineteenth-century geographical practice was rooted. The British geographers who mapped the interiors of colonial holdings in Africa and India during this period were considered heroic figures, and their discoveries set the agenda for exploration and exploitation of the resources of those lands.[44] The Martian cartographer who discovered and described new features likewise claimed a heroic position in planetary astronomy.

### HEROIC ASTRONOMY GOES WEST

When American astronomers took up the question of life on Mars, the heroic model was transformed by its encounter with the myth of the frontier. This occurred first because of the location of the new American observatories, built by wealthy industrialists atop remote mountains in the American West. The association of observatories with the isolated landscapes of the West, combined with ideas of frontier manliness, helped define the new persona of the American "astronomer-adventurer."[45] As remote as they may have been from the urban centers of the East Coast, they were connected by new technologies, like the telegraph, that made such distances less isolating. They were able to combine the authority of the astronomer-adventurer and his big telescope under clear skies with the immediacy of electronic and mass-media print communication. Dispatches from these outposts became popular fodder for the burgeoning newspaper industry and the new journalism of the late nineteenth century. Mars oppositions now became highly anticipated events, as readers awaited news from the observatories about life on Mars.[46]

One of the first American astronomers to tie his observational adventures to the news cycle in this way was William Pickering (1858–1938). His first exploit was as the leader of a Harvard University eclipse expedition in northern California. The reports Pickering sent back by telegram read like an explorer's dispatches, their adventurous tone "tempered by a cool scientific rigor that artfully drew attention to the technoscientific systems that constituted the event as a unique, difficult, and special enterprise."[47] Pickering followed up these exploits with an ambitious expedition to Arequipa, Peru, again on behalf of Harvard University. While the aim of this adventure was not meant to be the observation of Mars, Pickering could not help but take advantage of the Southern Hemisphere's unique view of that year's opposition, which was best viewed from a southern latitude and which, due to the fact it was perihelic (when Mars is closest to the Sun in its orbit), promised to be one of the best opportunities for witnessing new Martian features since Schiaparelli had first reported the *canali* fifteen years earlier.

Pickering had suggested in 1888 that Schiaparelli's canals were visible not because of their great width, but because of adjacent strips of vegetation that the canals fed during the icy poles' seasonal melting. Like Proctor before him, Pickering saw no reason not to assume that Mars harbored at least plant life. During the 1892 opposition he reported that his theory was confirmed by observation. He also reported seeing forty large lakes, as well as clouds in the atmosphere. Pickering saw himself as following in the tradition of the great explorers; his dispatches to the press and his letters home read as though he had just "returned from an unknown land, teeming with new revelations."[48] In one letter to his mother, he wrote, "I have been very busy upon Mars this week, I begin to feel quite at home upon the planet, and feel as if I could quite find my way about upon it anywhere." He looked forward to his nightly visits to Mars and its "countries."[49]

Pickering's view of Mars, as well as his model of communicating news from Mars, influenced the work of the man whose name is today synonymous with the idea of an inhabited Mars: Percival Lowell (1855–1916). Lowell was well acquainted with Western appetites for sensational information about foreign lands. When he first heard about Schiaparelli's provocative reports, he was in Japan, having made a career for himself as a travel writer with a flair for orientalism.[50] But one of Lowell's first loves was the nebular hypothesis, and the news that astronomy was discovering an inhabited Mars rekindled his desire to write treatises about cosmic philosophy.[51] After returning to the United States in 1893, he announced that he was financing a Martian expedition to the Arizona Territory. William Pickering, whom Lowell enlisted to assist in this endeavor, no doubt helped to stoke Lowell's Mars enthusiasm and influenced his thinking. By 1894, Lowell had decided that he would build an observatory in northern Arizona devoted to the investigation of life on other worlds.[52]

Ensconced in his Flagstaff observatory, Lowell became a "hybrid astronomer-author-adventurer on the Colorado Plateau."[53] His theory about the Martian canals evolved rapidly, and was fully formed by 1895. Less cautious than Schiaparelli, Proctor, or Pickering before him, Lowell claimed that Mars was not just habitable, but inhabited by intelligent beings who had covered the planet in a complex system of canals to survive on what had become a desert world.[54] Lowell took Schiaparelli and Pickering's Mars, with its oceans and lakes, and turned it into a vast, dry wasteland with only sparse vegetation. The geometric nature of the canals was all the visual proof Lowell felt he needed for his claim that the canals were artificial: "If Dame Nature be at the bottom of it," he wrote, "she shows on Mars a genius for civil engineering quite foreign to the disregard for prosaic economy with which she is content to work on our own work-a-day world."[55] The canals were straight, the apparent vegetation that bordered them was of uniform width, and their arrangement seemed to be coordinated with features Lowell took to be reservoirs.

Beyond visual evidence, Lowell drew upon his own version of the nebular hypothesis as it applied to the evolution of the planets. Every planetary orb in Lowell's model went through six stages. First was the sun stage, in which the planet was hot enough to radiate light; second was the molten stage, in which the planet was still incredibly hot, but beginning to cool; third was the solidifying stage, in which the planet developed a solid surface; fourth was the terraqueous stage, where there was liquid water at the surface; fifth was the terrestrial stage, when the oceans disappeared; and sixth was the dead stage, with neither water nor atmosphere. Mars, because it was smaller than Earth, had already reached the terrestrial stage. The Moon, smaller than Mars, was what the red planet would become once its water and atmosphere were fully depleted. Mars was thus not the twin of Earth, but an elder sibling and an example of the next step in Earth's own evolution, as all planets were at the mercy of the same set of natural forces.[56]

We should also take into account the context in which Lowell put forth this discussion of canals and civil engineering. At the end of the nineteenth century, canals were seen as symbols of progress. To make canal building the ultimate project of a superior race of intelligent Martians was to celebrate the achievement of the industrious builders of the Suez Canal, opened in 1869, and the Panama Canal, still under construction.[57] These canals allowed nations to assert global power in not just navigating the globe, but cutting through land masses that presented inconvenient obstacles. They were heralded as evidence of the mastery of natural landscapes, industrial forces, and world politics. A similar comparison could be made between the Martian hydrological system and colonial engineering projects of the era that were aimed at water management and desert reclamation.[58] This optimistic interpretation of the canals was echoed in Boston *Transcript* editor Edward Henry Clement's poem "The Gospel from Mars." The poem celebrated

the canal builders of Earth as the day's "conquering heroes," while describing how their counterparts on Mars similarly fought back the planetary drought and encroaching deserts of their world. The poem was an
ode to the power of engineering.[59]

Technological optimism notwithstanding, the canals tapped into a
general anxiety about desertification, as the dry desert landscapes described by Lowell resembled "the terrifying droughts in India, South
America, and China from 1877 to the end of the century."[60] Lowell certainly was of the belief that terrestrial deserts were only bound to increase. As he described them, they were evidence of the "unspeakable
death-grip" of the later stages of planetary evolution. Deserts already
existed on Earth, and experiencing them gave one a horrible appreci-
ation for what was to come: "For the cosmic circumstance about them
which is most terrible is not that deserts are, but that deserts have begun to be."[61] Deserts, and the occurrence of droughts on Earth, marked
for Lowell the beginning of the end of Earth's habitability. His vision
of intelligent Martian engineers was thus not ultimately an optimistic
one. It was also not incredibly progressive. Lowell insisted that such a
massive engineering project was only possible in a well-ordered hierarchical society. As a dedicated social Darwinist, he believed, like Chambers before him, that such a stratified society was the natural result of
human evolution.

### THE END OF MAPPING CANALS

The Greek-French astronomer Eugene Michel Antoniadi (1870–1944)
settled the canal debate in the early twentieth century. He observed the
perihelic opposition of 1909 for two months using a thirty-three-inch
refractor at Meudon, the largest telescope in Europe at the time. Antoniadi had by that point been director of the British Astronomical Association's Mars Observing Section for several years, and had developed
his own criteria for evaluating and weighting the observations of his
fellow section members. Along with his position came a great deal of
institutional authority with which to make claims about Mars. While
he had started out believing in the artificiality of the canals, his experi-

ence led him to question whether these features weren't in fact beyond the resolving power of the telescope. He set out to determine once and for all the true nature of these features.

Rather than attempt to map the planet or add new features, Antoniadi instead performed a close examination of a number of the canals proposed by Lowell and other astronomers. In the end he determined that "The geometrical canal network is an optical illusion; and in its place the great refractor shows myriads of marbled and chequered objective fields, which no artist could ever think of drawing."[62] By "optical illusion," Antoniadi did not mean that the canals were not based on some real feature of Mars. As he explained in a letter to Lowell, he believed the network of canals to be an "optical symbol of a more complex structure of the Martian deserts, whose appearance is quite irregular to my eye." He told Lowell that he did see some canals, and even interlacing canals, but that they were "a vague network," and "irregular and knotted or diffuse."[63] In other words, the canals that did resolve were natural features.

Antoniadi did ultimately produce his own map of Mars, based on his sketches from multiple oppositions. As we might predict, it did not include a recognizable network of canals; in their place were complicated and mostly indecipherable surface details. However, Antoniadi remained convinced that there was life on Mars. In his scientific report he wrote, "We thus see in the so-called 'canals' a work of Nature, not the intellect; the spots relieving the gloom of a wilderness, and not the titanic productions of supernatural beings." There was no great engineering project at work on Mars, only "the natural agencies of vegetation, water, cloud, and inevitable differences of colour in a desert region."[64] Antoniadi still believed that it was quite probable that animals and even humanlike creatures could have existed on Mars. However, agreeing with Lowell that Mars was a dying world covered in dry deserts, he concluded that "advanced life must have been confined to the past, when there was more water on Mars than there is now."[65] This is the Mars that the early space age inherited. Antoniadi's work, summarized in his 1930 book *La planète Mars*, was the last word on the topic of canals.[66]

## MARS LITERATURE COMES OF AGE

If the imagined journeys of the previous chapters were more mysterious than otherworldly, the cosmic voyage literature of the eighteenth century made up for this by presenting a universe teeming with forms of life often as familiar as they were alien. These journeys remained focused on theological matters, with the added impulse of exploring the philosophical and religious implications of a plurality of inhabited worlds. These intellectual musings overlapped with concerns related to the encounters with other cultures that accompanied the Second Age of Discovery, as Europeans attempted to understand themselves as part of a larger human family. As such, this literature is replete with representations of the noble savage and the colonized subject.[67] The Moon remained the most popular destination for cosmic travelers in the Enlightenment, but some did visit Mars. As astronomy had not yet provided any topographical guidance, the authors who attempted descriptions of Mars and its inhabitants prior to the nineteenth century presented an unconstrained variety of ideas about what the red planet might be like. A few examples will illustrate just how different Mars could be, as well as how it was shaped by the questions one brought to the intellectual exercise. After this, we will examine the trajectory of Mars literature as the living surface of Mars seemed to come into focus.

In Eberhard Christian Kindermann's 1744 novel *Geschwinde Reise auf dem Luft-Schiff nach der obern Welt* (An airship journey to the upper world), a group of Christian travelers visit a fictional Martian moon. Here they encounter an Edenic world of lush and colorful vegetation, populated by mythological creatures and peaceful humans. The travelers are at first worried about what the presence of other humans in the solar system means for their own place in God's plan. However, they soon conclude that these humans live outside of Earth's salvation narrative.[68] Bernardin de St.-Pierre's *Harmonies de la nature*, published serially from 1792 to 1814, imagined that Earth held God's template for all life in the solar system, and that each planet might be similar to a region of Earth. He distributed the climates and races of his home planet into the cosmos. Mars, as he described it, was like Earth's cold polar regions; it was populated by sea lions, whales, and humans reminiscent of northern Germans or Scandinavians.[69]

The Yorkshire vicar Miles Wilson in 1757 imagined the convoluted story of a 1,700-year-old man who survives and suffers because Christ has told him to walk the Earth until the apocalypse. Wilson's novel begins as the author discovers and translates from Chinese into English the memoirs of Israel Jobson, the "Wandering Jew." Intended as a primer on astronomy through a Christian lens, Jobson's fictional memoir recounts a period spent visiting the planets in the company of an angelic guide. Jobson of course visits Mars, where he discovers nine million red, sexless intelligent beings growing like trees, rooted in the Martian soil with their eyes pointed up toward the heavens. The angel tells Jobson that these beings, who have been growing in place since the planet's creation, are yet of no significance in the Christian narrative; they will ultimately be consumed when the planets at last fall into the Sun: "They will be dissolved at the Conflagration as being Lumber not fit for the Celestial Mansions."[70]

Mars fiction changed dramatically in the late nineteenth century as its authors began to employ the same theories of celestial and biological evolution that influenced astronomers and authors like Chambers. Thanks to maps and spectroscopic evidence, the landscape of Mars also solidified during this period, first into a land of seas and continents, and then into one of canals and intelligent beings. The volume of Mars literature also increased during this period as new printing technologies and increased literacy made it possible to produce cheap novels and magazines for readers interested in more speculative ways to explore the implications of the new science. The number of space fiction novels printed from the late nineteenth century to first quarter of the twentieth "is barely measurable."[71] The new mass market was flooded with stories about Mars, which, thanks to the publicity surrounding the astronomical observations of the red planet described above, displaced the Moon as the destination of choice for fictional space travelers.

One notable addition to space fiction during this era was something resembling the modern spaceship. While few authors yet imagined anything like the rockets and capsules that would take humans into space in the late twentieth century, they nonetheless imagined enclosed vessels that could keep humans safe during their travels, and which could provide many of the comforts of home (in this way they were very similar to the spaceships we still imagine are in our future). As for the problem of how to escape Earth's gravitational pull, many of the fictional travelers describe inventing versions of antigravity technology: harnessing secret forces in the universe or newly discovered physical substances that allow them to travel effortlessly from one world to another. The ship Percy Greg devised for his 1880 journey to Mars, *Across the Zodiac*, was called the *Astronaut*, possibly introducing this word into English for the first time. The *Astronaut* is described in the terms of a large sailing vessel, and its chambers include libraries, gardens, and even an aviary. It travels via an antigravitational energy that Greg dubs "apergy."[72]

Fictional works like Greg's helped readers to witness Mars through their mind's eye by presenting them with accounts from travelers who had experienced Mars's surface and inhabitants firsthand. Late-nineteenth-century authors treated Mars not as a mystery but as a solid and constant world they constructed from their understanding of astronomy, evolution, and society—effectively taking part in, or at least latching onto, astronomical discoveries and debates.[73] These fictional accounts helped Mars to emerge more fully formed, and endowed with meaning and cultural significance.[74]

One impulse in these turn-of-the-century Mars fictions is to imagine Mars in either utopian or dystopian terms. In these stories the planet becomes a lens through which to envision what an ideal society might look like, or celebrate Victorian ideals as utopian.[75] Greg imagined his hero's journey in the wake of Schiaparelli's discovery of the *canali*, and even before Lowell's insistence that these were proof of intelligent life on Mars, he was able to speculate, based on the nebular hypothesis and evolution, that Mars might be populated by intelligent beings like ourselves. Rather than being superior life-forms, however, Greg's Martians are joyless and overcultivated beings whose lives are made too easy by the mechanization of labor and the hyper-rationalization of politics and social life. Recognizing these Martians as communists, Greg's protagonist joins up with an underground society of Martians bent on overthrowing the government and returning the planet to more suitable (i.e., Victorian) ideals.

H. G. Wells's 1898 novel *The War of the Worlds*, with its horrific firsthand account of a Martian invasion of Earth, remains one of the most famous Mars stories of this period. Wells followed astronomical debates about Mars's habitability with great interest. He used the novel as an opportunity to explore the idea that Martians, having evolved on a world much older than Earth and perhaps more hostile, might be something other than more perfect versions of human beings. They would, he figured, be truly and completely Martian. They might not even recognize humans as intelligent beings like themselves. Critics of the novel could find no fault with Wells's scientific or philosophical approach to imagining Martian life and civilization, based as it was in well-known observations and theories about Mars, as well as evolutionary theory he had learned directly from Charles Darwin's chief disciple, Thomas Henry Huxley. Indeed, the first pages of the novel contain a detailed summary of scientific thinking about Mars, as well as a thorough account of Martian observations (up to Schiaparelli, but excluding Lowell).

Unlike Greg, Wells used Mars not to celebrate Victorian ideals but to skewer them. Wells's Martians are superior to humans, but for all their intelligence and technological prowess they are monstrous and without sympathy. They are huge slimy octopods who feed on human fluids like alien vampires, and their technologies are designed to conquer and exploit the local resources, humans included (plate 4). Wells held this image of the merciless Martian, shaped by evolution and circumstance, up to that of the British Empire, whose power came from its military might, its presumed superiority, and the bloody grasp it held on its farflung colonies. Wasn't Britain guilty of these same crimes against humanity? Had Europeans not exterminated animals and peoples of Earth with the same abandon?[76] In the end it is the Martians' own biology—

their lack of defenses against Earth's bacteria—that thwarts their initially unimpeded invasion of the planet.

The year before Wells's novel appeared on bookshelves, when it was still appearing in serialized installments, the German author Kurd Lasswitz produced his own influential story of Martian invasion. *Auf zwei Planeten* (On two worlds) had aims similar to those of *The War of the Worlds*; it sought to hold a mirror up to humanity, and make readers examine their complacency and presumptions of superiority. The Martians whom Lasswitz imagines, however, are very different from Wells's merciless killers, though no less dangerous. Lasswitz's novel begins with a balloon expedition to the North Pole. When the two adventurer-heroes of the novel are blown off course, they encounter a group of humanoid Martians—or, as they identify themselves, Nume from the planet Nu—who have surreptitiously installed a wheel-shaped space station above the pole. Here the human adventurers learn that the Martians seem to have arrived as benevolent colonizers, bringing with them the solar-powered technology they use to produce food directly from Earth's raw materials. On their home world, this same technology has ended scarcity and produced peace. The two heroes are brought to Mars and given a tour, where they see how the Martians do live peacefully and in harmony with their planet.

But Lasswitz's Martians also have imperial ambitions. Their antigravity technology has led them to think of themselves as "the masters of the solar system."[77] The Martians covet Earth's resources, its abundant supply of water, and its closer proximity to the Sun. Here the plot begins to highlight how the colonizing nations of Earth might respond to similar actions taken against them. The British Empire does not take kindly at all to these enlightened invaders, and attempts to fight them off. In response, the Martians establish Earth as an occupied protectorate. The Martian occupiers are eventually ousted, and peace is established when a clandestine American operation learns to pilot the Martian ships and builds a secret fleet in Central America to strike against the polar station.

Mark Wicks was one of many authors at the turn of the century who sought to align his fiction directly with Lowell's claims. An amateur astronomer himself, Wicks was in correspondence with Lowell when he started writing his 1911 novel *To Mars via the Moon*. The story is narrated by a man named Wilfrid Poynders, who like the author is an amateur astronomer fascinated by the planet Mars. Poynders and two other men build a ship they name the *Areonal* from a special alloy called Martalium, to carry them to Mars. Poynders, who is mostly referred to throughout the novel as "Professor," provides the funds and the astronomical knowledge, while John Claxton, an electrical engineer obsessed with airplanes and other flying machines, and Kenneth M'Allister, an experienced naval mechanic and engineer, provide much of the practical know-how essential for building and operating the vessel. The book occupies a space somewhere between interplanetary romance and popular science text, as its action is regularly interrupted by Socratic dialogues and lectures in which the Professor answers his crewmates' questions. The names of several astronomers are invoked throughout, as the Professor provides a full exploration of observations and theories about the Moon and Mars during the journey.

Perhaps to a greater degree than the texts already described, *To Mars via the Moon* also allows the reader to explore and witness the Moon and Mars via its protagonists. The Mars at which the *Aeronal* arrives is exactly as Lowell described, and the Martians the travelers meet are technologically and culturally advanced. Wicks is also able to go beyond Lowell and combine evolution with the popular spiritualism of the time; the novel illustrates that evolution is not random, but is guided by a benevolent power that allows Poynders to be reunited on Mars with his reincarnated son.[78] Wicks uses Mars to show his readers what progress along the lines established in the Victorian era will bring: the Martians know no poverty, no war, and no discord whatsoever.

### THE FINAL FRONTIER FANTASY OF MARS

Such intellectual exercises would not survive the twentieth century. The American pulp science fiction and fantasy magazines would transform Mars from a philosophical and scientific thought experiment into a backdrop for swashbuckling adventure. In the pages of the pulps, the idea of a sophisticated Martian civilization was replaced by dry and dusty ancient towns ruled by pirate kings and warlords of various Martian species. Sam Moskowitz, in his short history of pulp Mars science fiction, credits one author in particular with "turn[ing] the entire direction of science fiction from prophecy and sociology to romantic adventure."[79] That author was Edgar Rice Burroughs. His Mars stories took place on a planet the local inhabitants called Barsoom, and featured the violent exploits of an Earthling named John Carter.

For Burroughs, Mars was an extension of the American frontier, and a setting for recreations of the mythology of the Wild West. He relocated and reconfigured the disappearing frontier of the American West on Mars. The final serialized installment of his first Barsoom novel, *A Princess of Mars*, appeared in the same year that Arizona became a state. The novel is set in the 1860s, shortly after the end of the American Civil War, with Carter a former Confederate captain and plantation owner dispossessed of his land and property. That Carter, now a prospector, travels mysteriously between the Arizona desert to the surface of Mars denotes a cosmic link between the two places and their respective mythologies of civilization at the end of a sword. Burroughs's popular series of Barsoom adventures made visions of the red planet "as much recreation of the past as a vision of the future." They did not question or critique Southern masculinity, westward expansion, or even slavery. If anything, they celebrated those things as they mourned the disappearance of the frontier and mythologized the culture and values of the antebellum South. They presented "neo-Arthurian codes of honor," chivalrous violence, romance, and revenge.[80]

These Martian adventures wouldn't last forever, but they did outlive Lowell's intelligent Martians, even as Antoniadi's version of Mars came to dominate planetary astronomy. This early Mars science fiction, even with its flaws, was influential. It inspired the rocket engineers and explorers of Mars whom we will meet in the next chapter. Wernher von Braun, who built the deadly V-2 missile for the Nazi war effort and then the American rockets that took the first humans to the Moon, was particularly influenced by Lasswitz's work. Von Braun even tried his own hand at imagining a fictional Mars expedition, in which he presented his concepts of the technologies and technocratic management required to make such a vision into a reality. As we will see, such ideas among the founders of the American and Soviet space programs made human Mars exploration seem like the logical next step after the Moon. Forces beyond their control—including Mars itself, which refused to live up to expectations but insisted instead on becoming another planet altogether—conspired to keep Mars exploration a robotic endeavor.

# COLD WAR RED PLANET

In the space age, planetary exploration via robotic space technologies began to constrain the possibilities of life on Mars. The red planet went from being an extreme desert version of Earth to being a planet that a seemingly separate evolutionary path had rendered still and lifeless. This happened in the context of an increasingly militarized world in which two global superpowers invested in missile technologies that made spaceflight possible. Rockets were developed as weapons, but success in space became its own form of high politics. Launching space vehicles legitimized claims about missile power and technological capabilities, and dominating space became a proxy for geopolitical power. The competition between the United States and the Soviet Union in space provided the impetus for the first era of human and robotic exploration of the Moon and the planets.

The Cold War was global in nature. It involved missiles that could travel higher, faster, and farther than ever before; nuclear bombs that could destroy entire cities in an instant; and nuclear-armed submarines that could clandestinely traverse the ocean depths. At the end of World War II, from a military standpoint, knowledge of the new potential weapons and theaters of war was underdeveloped. Employing these new technologies, not to mention protecting against them, required not only technological advances but a new science of Earth and the planets.[1] It was unclear exactly what knowledge of these relatively unexplored regions would be of strategic military value. This certainty that more needed to be known, together with uncertainty about exactly what knowledge was needed, led to new government and military patronage for science, and a new relationship between the scientists and their patrons.

In the United States, the technologies and funding opportunities made available by the state shaped the development of planetary science. In supporting civilian spaceflight, the government subsidized "complex feats of intellectual labor" and built a visible and captive community of experts with useful skills working in the national interest.[2] Scientists who wished to ask questions about Mars had to learn how to ask them in the new technical language of spaceflight. To design instruments for Mars exploration, they worked with NASA engineers, mission planners, and the same aerospace contractors who were engaged in solving problems for the military and the CIA.[3] They worked within constraints and timelines that were often set with little regard for science. Many of the scientists who participated in the early years of rocket-powered science were those who were willing to tinker and invent within "a tool-building technical culture."[4] This community was cultivated for its usefulness in the continued development of space technologies that interested the state.

### EXPLORING MARS WITH MARINER

It would be nearly impossible to tell the story of US robotic Mars exploration without also telling the story of the Jet Propulsion Laboratory (JPL), the center that led in development of interplanetary spacecraft during the Cold War. Formerly a wartime center for experimental military rocketry, JPL first caught the public's attention in 1958 after the launch of America's first successful artificial satellite, Explorer 1, on January 31, 1958. One year later, the lab became part of the newly formed National Aeronautics and Space Administration (NASA), where it was assigned responsibility for robotic exploration. As part of the space agency's first ten-year plan, JPL was directed to develop a series of lunar missions called Ranger and Surveyor—named to present lunar exploration as a continuation of frontier exploration and mapping. They would also develop a series of Mariner probes to be sent to Mars, Venus, and Mercury and, ultimately, a robotic soft-lander called Viking. The names Mariner and Viking evoked the great historical voyages of discovery that had brought Europeans to the shores of unknown worlds.[5] These missions were ambitious, and their timelines quickly outpaced the laboratory's wartime "shoot and hope" model of technology development.[6]

The second Mariner mission, which flew by Venus in 1962, was the lab's first real success. Writing up a preface to the official Mariner 2 mission history, JPL's director, William Pickering (unrelated to the other William Pickering who helped Percival Lowell build his observatory, as explained in the last chapter), announced that the Mariner program had opened up "a whole new era of experimental astronomy." It was a new science that required "many thousands of man-hours" for design and operation of "complex automatic equipment which must operate perfectly in the harsh environment of space."[7] Such an endeavor was challenging, but Pickering was confident that JPL, its industrial contractors, and its collaborating scientists—a team that he estimated numbered "several thousand men and women"—had only just begun to taste victory. The report announced that "Mariner was exploring the future," allowing humans at last to find answers to long-standing questions about their place in the universe.[8] This exploration would be done not by the hero-astronomers of the nineteenth century, but by large teams of scientists and engineers.

For those interested in the search for life, what Mariner 2 found at Venus shifted attention more fully to Mars. Venus was covered in a dense, cloudy atmosphere and had day and night temperatures hot enough to melt lead: an inhospitable nine hundred degrees Fahrenheit. Mars, on the other hand, still seemed to hold hope for life. Ground-based observations of Mars using non-visible spectra suggested that the planet was incredibly arid, but not necessarily bereft of all life. Some experts, like the Soviet astronomer Gavriil Adrianovich Tikhov (1875–1960), suggested that something similar to the lichens that can be found in extremely cold and dry environments on Earth might thrive on Mars.[9]

The spectra the astronomers collected could be highly suggestive, leading some to believe that plant life on Mars had in fact been confirmed. Looking at Mars in the infrared during the 1956 opposition, the astronomer William Sinton (1925–2004) thought he had found the spectral signature of something that could be lichen (or lichenlike). In 1958 he continued his measurements, this time paying close attention to the dark and light regions of Mars's surface. Finding that the dark regions produced spectra similar to those of the large organic molecules and carbohydrates associated with plant photosynthesis, Sinton concluded that the dark regions were green vegetation.[10] Experiments done in US Air Force laboratories by the German-born physiologist Hubertus Strughold—a former Nazi scientist, though not a member of the Nazi party—seemed to confirm that lichens could grow in Martian conditions.[11] Plant life on Earth existed in extreme environments, and evolutionary biology had shown life to be incredibly adaptable. Perhaps, as one 1958 article argued, the Martian plants had even evolved "as a metabolic by-product some kind of antifreeze."[12]

Four spacecraft made the journey to the red planet as part of the Mariner program. Mars received its first flyby visitor in July 1965—four years before the first human set foot on the Moon—when Mariner 4 made its closest approach to the planet. Mariner 4 obtained twenty-one images of Mars (and a twenty-second partial image) from distances ranging from ten thousand to seventeen thousand miles. These images gave scientists a grainy but detailed glimpse of roughly 1 percent of Mars's surface, at distances from which even the largest rivers on Earth would not be visible.[13] The images showed what looked like an ancient and cratered surface, much like what the Ranger 7 lunar spacecraft had shown of the Moon a year earlier (figure 9). The imaging team interpreted the images as showing a planet that was more Moonlike than Earthlike, but with a thin atmosphere. Because of the number of craters seen in the images, the team assumed that the surface was billions of

FIGURE 9 Mariner 4 flyby image of Mariner crater. NASA/JPL.

years old, unchanged by internal processes such as volcanism, and un-weathered by water. Magnetometer readings indicated that the planet had no magnetic field. Mars, it seemed, was a dead planet—at least from a geological point of view.

The question of whether life existed on Mars, or had existed there in the past, was still open. So too was the question of whether or not Mars held clues to the prebiotic history of Earth.[14] Despite the seemingly bleak prospect of finding life on a Moonlike planet, the US National Academy of Sciences Space Science Board argued that searching for life on another world would mark a milestone in human history, and its potential consequences were so great that it should be given the highest priority.[15] Meanwhile, NASA's own Ad Hoc Science Advisory Committee strongly suggested in 1966 that "the time is surely here when we must define minimum success in terms not only of 'getting there,' but in terms of scientific accomplishments."[16] Not all NASA programs should be oriented toward Cold War competition; scientific objectives should be given more priority. The search for life fit this bill.

Two more robotic flybys of Mars followed, with Mariners 6 and 7 launched during the opposition of 1969. They encountered Mars less than two weeks after Apollo 11 landed the first humans on the Moon. The twin spacecraft were improved versions of Mariner 4. The imaging system had improved resolution and was attached to a movable platform that allowed camera pointing.[17] The two craft were able to collect nearly ten times as many images as Mariner 4—a total of 201 images. In addition, the spacecraft carried more instruments. The television imaging system was supplemented by instruments that could gather data in infrared and ultraviolent wavelengths. The onboard computer could handle up to thirty-five times more data, and could also be re-programmed. Earlier Mariner computers had been relatively simple automated sequencers that could perform preprogrammed routines on command and relied on electronic clocks or radio signals from Earth to initiate actions.

The new Mars of Mariners 6 and 7 was less Moonlike; some areas showed very few craters. But it was still far from Earthlike. Mars also had features unknown on either Earth or the Moon, such as large regions of chaotic terrain. It had ravines and crevasses. These new findings "moved the conception of the planet into an anomalous position. It was neither Moon nor Earth, and it had its own unique properties."[18] But scientists had still only seen a very small percentage of the planet's surface.

### IMAGINING HUMANS ON MARS: AN APOLLO PROGRAM FOR THE RED PLANET?

The Mariners and the Earthbound explorers who traveled vicariously with them to Mars were asking fundamental questions about their universe and the place of life within it, but they were also Cold War heroes. They represented a growing community of scientific and technical experts who enjoyed the patronage of a government willing to spare little expense in developing the knowledge and know-how necessary to dominate an unpredictable global battleground. As the United States and the Soviet Union raced to explore Mars and Venus, every success for JPL in accomplishing an interplanetary first was rewarded. While the Soviet space program's robotic probes captured many of these firsts at the Moon, JPL gave the United States a strong claim on the red planet. President Lyndon Johnson gave the Mariner 4 team medals for their work, met personally with a select group of team members in the Oval Office, and listened to their explanations of the Mars images returned.[19]

Government support had its limits. It encouraged increasingly complex and sophisticated missions, allowing the return of great troves of scientific data. The study of craters on the Moon and Mars overlapped with similar studies at missile test ranges, and spacecraft methods of understanding the geology of Mars could be applied to satellite reconnaissance of terrestrial spaces.[20] Expertise in designing autonomous spacecraft with advanced navigation and tracking systems that could survive radiation hazards and high-velocity impacts likewise had military and intelligence value.[21] For a brief time, the US Air Force even considered Mars as a possible site for military operations, and supported research into the challenges of living on Mars.[22] But for those who hoped for a human mission to Mars to follow Apollo, the prospects were becoming increasingly dim. In the background of the development of Mariners 6 and 7, NASA was struggling to define its future. Even as Apollo was on the eve of making good on its promised goal of sending humans to the Moon, the agency's budget was in decline.

For those who wanted big-budget Apollo-style exploration to continue, Mars seemed like a natural choice. To send humans to the red planet would first require better knowledge of the planet's surface. NASA proposed a large, robotic, life-seeking lander called Voyager that would launch atop the Apollo program's Saturn 5 rocket—the largest and most expensive rocket in the US arsenal. For NASA Administrator James Webb, this and the human missions that would follow would help maintain NASA's budget, provide justification for continuing to produce the powerful rockets, and extend NASA's mission. But while

Voyager initially garnered support from Congress, by 1967 that support had waned.

Mars Voyager was expensive. Webb needed a more affordable program he could sell to the White House. He came back with a scaled back lander called Viking that could be launched on a smaller and more affordable rocket. As the Soviet Union was planning its own Mars lander missions, Webb felt a cheaper lander option had legs.[23] He sensed a change in the political landscape, and was careful not to present Viking as a foot in the door to human exploration of Mars. Webb left NASA in 1968, leaving the agency with no clear post-Apollo human spaceflight mission but several small robotic programs and hopes for a way forward in human spaceflight.

Not everyone gave up on human missions to Mars. When Webb's deputy, Tom Paine, took over as NASA administrator, he sought to promote Viking as a precursor to human missions. In 1969 and 1970, Paine attempted to sell an ambitious post-Apollo program to the newly inaugurated Nixon White House, in which Viking would be a "first step in a series of robotic flights culminating in human exploration of the Red Planet by the end of the twentieth century."[24] He proposed that Mars exploration would function "in an Apollo mode, with a strong headquarters director making use of multiple NASA centers, industry, and universities."[25] Paine received some encouragement from the words of Vice President Spiro Agnew, who presented Mars as a new frontier, the conquest of which would usher in a new age of discovery and "a new era of civilization."[26] As the Apollo 11 landing took place in the new administration's first year, Paine wanted to capitalize on this triumph and use it to push for a new American goal of putting humans on Mars as soon as possible—as early as the 1980s.

To draw up plans for a human Mars program, Paine called in the former Nazi rocket engineer turned American Cold Warrior, Wernher von Braun. Von Braun had been thinking about this topic since at least the 1940s, if not earlier. In 1947 he had begun writing a technically detailed science fiction novel, *Das Marsprojekt* (*The Mars Project*), that he hoped would raise enthusiasm for spaceflight. This was "an elaborate spaceflight feasibility study in the form of a science fiction novel,"[27] and it was pure Cold War techno-fantasy. It imagined that a third world war had led to the defeat of the Soviet Union by the United States and its allies, and that space had been key to victory—specifically, a military lunar station that dropped atomic weapons on Soviet ground-based facilities.[28]

The novel's vision of a journey to Mars is grand—seventy men (and no women) travel to the planet aboard ten ships for a three-year round-trip voyage that includes time spent exploring the surface. Just assembling and fueling the ten ships in Earth orbit requires 950 separate launches over eight months. Von Braun estimated that the cost of such a mission would be two billion dollars.[29] The novel itself was not initially published, but its scientific appendix appeared in the 1950s. A scaled-back version of von Braun's Mars mission concept made its way into his popular magazine articles and onto Walt Disney's popular television programs *Man in Space*, *Man and the Moon*, and *Mars and Beyond*, in which von Braun speculated about the use of atomic-electric powered spaceships to reach Mars. Together, von Braun, Disney, and other popularizers influenced generations of Americans in imagining what a journey to Mars would entail.

In the 1960s, von Braun was given the chance to make his Mars fantasy a reality. At Paine's behest, he presented a Mars program to Nixon's Space Task Group (STG). The new plan envisioned a 1981 crewed mission to Mars that would involve a 640-day round trip, and which would be based on a series of precursor missions that would keep the NASA budget at roughly Apollo levels. The STG was convinced of the feasibility of the program, but cautious about the commitment. In drafting its 1969 report to the president, the STG presented a variety of options for NASA's future. It was frontloaded with the conclusion that "NASA has the demonstrated organizational competence and technology base, by virtue of the Apollo success and other achievements, to carry out a successful program to land man on Mars within 15 years." Human missions to Mars should be a long-range goal of NASA, it said, even though the necessary precursor missions could be carried out "without developments specific to a Manned Mars Mission."[30] For Paine and Agnew, sending humans to Mars promised "a new banner to be hoisted" to rally NASA's human spaceflight program and provide the next Apollo.[31]

The president wasn't interested in a new Apollo. Paine's vision collided with Nixon's space doctrine, which placed "critical problems here on this planet" on an equal footing with space priorities. Nixon wanted a less ambitious space program, not a continuation of Apollo-level funding and activity. Congress likewise balked at the idea of funding another program at Apollo levels. Paine managed to keep Viking and a supporting orbiter mission on the books, though the lander had to be postponed to a 1975 launch. He resigned in 1970, frustrated that he had not made Mars his legacy. NASA was in turmoil for the next two years as it struggled to define its priorities for the post-Apollo future. In 1972, NASA and Nixon finally agreed that the space shuttle and an orbiting station would be the agency's primary goals in human spaceflight, while planetary exploration would continue robotically. NASA stabilized.[32]

### MARINER 9

Human missions to Mars were off the table, but two more robotic missions flew to Mars during this first era of exploration. These would reveal a Mars very different from the one suggested by the previous flyby missions. Mariner 9 was the first mission to go into orbit around Mars, beating a rival Soviet orbiter by a matter of days. It was similar to the previous Mariners, but with an additional propulsion system for orbital insertion and a slightly better suite of scientific instruments than those carried by Mariners 6 and 7. In addition to infrared and ultraviolet spectrometers, the spacecraft also carried a radiometer that would search the planet's surface for hotspots that might indicate volcanic activity. With its closer proximity to the planet and a more advanced imaging system, Mariner 9 would be able to photograph Mars at resolutions up to eight times greater than the previous flyby missions. Since it would orbit the planet for an extended period, it would be able to map a large percentage of Mars's surface.

Fortunately for the scientists and engineers who had invested years of their careers in preparing for this mission, Mariner 9 also carried a new computer system built with a larger degree of in-flight adaptability. This came in handy first when its sister spacecraft, Mariner 8, was lost on launch. With the demise of the two-spacecraft mission, it was possible to reprogram Mariner 9 to accommodate some of the lost spacecraft's tasks. This adaptability came in handy again when Mariner 9 reached Mars on November 14. Earth-based observers watching Mars from observatory telescopes had noticed two months earlier that a bright yellow cloud was forming over the Noachis Terra region of southern Mars. By early October, the cloud had grown into a global dust storm. Over the next few weeks, the storm would become larger and thicker than any dust storm previously observed from Earth.

By the time Mariner 9 entered orbit, the storm was beginning to subside. But it raged on for another month. The spacecraft's mapping mission was delayed until the end of the year.[33] The Soviet Mars 2 and 3 orbiter team had no such luck; their spacecraft relied on a preprogrammed photography sequence and a film system. They used up all of their film before the dust abated, although they did manage to collect other data. The two spacecraft also carried automatic landers; the Mars 2 lander crashed into the surface due to a steep descent angle, while the Mars 3 lander made it to the surface successfully but operated for only twenty seconds before the instruments shut down. A third Soviet spacecraft bound for Mars failed to leave Earth orbit.[34]

Astronomers had observed Mars dust storms from Earth, but Mariner 9 allowed scientists to monitor a dust storm from closer than ever before. As they watched the dust levels drop slowly over that first month of the mission, surface features began to come into view. First to appear, along with the south polar cap, were four dark spots that the imaging team deduced were very high-altitude features unlike anything they had seen before. These turned out to be massive shield volcanoes—one of which, Olympus Mons, is the largest and tallest volcano humans have yet discovered (figure 10). It has a base the size of the state of Arizona, and is three times taller than the largest Hawaiian volcano, Mauna Loa.

After the mapping mission began, the science team was surprised

FIGURE 10 Apparent as a small, shadowy smudge near the top of this image, the giant volcano Nix Olympia, known today as Olympus Mons, showed itself to Mariner 9 as a global dust storm began to abate. NASA/JPL.

again to find an extensive canyon system five times as deep as the Grand Canyon, stretched over a distance equivalent to a road trip between Los Angeles and New York City.[35] They named the canyon Valles Marineris, to mark its discovery by the Mariner team. These exciting discoveries led the Mariner 9 team to conclude that only then had they actually started to see the planet. In the preface to the book of images the team published, they lamented that the earlier missions had returned images that only represented one type of Martian terrain. "It was almost as if spacecraft from some other civilization had flown by Earth and chanced to return pictures only of its oceans," they wrote.[36] The Mars that Mariner 9 mapped, said one team member, "was not the same planet for which this mission had been planned."[37]

After almost a full year mapping Mars, the spacecraft ran out of fuel and the mission ended, the team no longer able to control the space-craft's attitude. By then the Mariner 9 camera team had imaged the planet's entire surface and provided photographs that could then be hand-placed into mosaic maps and globes of the planet (figure 11). These maps wiped away the perception that Mars must be either Moonlike or Earthlike. The nineteenth-century idea of Mars as a frontier wasn't completely gone. But the language of the astronomer-adventurer was replaced by a new type of explorer: the field geologist. Geologist Thomas

**FIGURE 11** A photomosaic globe made from more than a thousand Mariner 9 images was the first global representation of Mars as we know it today. This globe is on display at the Smithsonian's National Air and Space Museum in Washington.

"Tim" Mutch used this language when he compared the daily image returns to "a new traverse across unfamiliar terrain." Mutch argued that each traverse revealed to the trained geological eye a previously unknown planet, wiping away memories of "the 'old' Mars as though the countless hours of previous speculation had been little more than science fiction."[38] The geologist on Mars was like the natural philosopher experiencing the tropics for the first time, realizing the folly of old philosophies.

This new Mars was perhaps not entirely dead. In addition to the conspicuously large volcanoes and canyons, images revealed channels possibly carved by water due to the seasonal melting of permafrost, and large dune fields that spoke to wind-driven processes eroding, depositing, and redistributing geologic material around the planet. The scientists attached to the mission tracked clouds of carbon dioxide and water ice, and were able to determine atmospheric circulation patterns for the first time. They observed the seasonal retreat of the polar ice caps. One of the most spectacular findings was the discovery of large channels that hinted at a history of massive flooding at some point in Mars's past.[39] Mars was interesting again.

The Mariner 9 science team came away with a new picture of Mars, and also new questions. Mars was a planet on which volcanism had played a major role, most visibly forming extensive plains and enormous mountains. But the absence of a magnetic field and the impressive height of the volcanoes indicated that the interior of the planet was probably not very dynamic, and lacked significant tectonic activity. The presence of Valles Marineris indicated that something significant must have happened in Mars's history to form such a complex feature, but there was no explanation for what that event might have been, or where the material that once must have filled the canyon had gone. The channels—some of which resembled river deltas, streambeds, runoff channels, and outwash plains—could allow one to imagine large amounts of water flowing on Mars's surface, but it was uncertain how that had happened, where the water had come from, and where it had gone.[40] Had the water flow happened all at once? Were the channels evidence of a massive flood generated by a catastrophic event that had suddenly warmed the planet and thickened its atmosphere? Had the flooding happened when Mars was a warmer, wetter world sometime in its past? Or was it possible that rare conditions could still occur on Mars that allowed liquid water to flow?

The geologists preached caution in drawing too many conclusions based on these features. It was time to perform laboratory experiments, go into wind tunnels, and do field experiments in analogous environments. It was time to figure out the physics of what had been observed, not to draw conclusions.[41] Mars now had a geologic past that could be interrogated. The planet also seemed at least somewhat more promising for the search for life than the early flyby missions had made it appear. Mars had probably been inhospitable to life for billions of years. But if conditions had been right for life early on, and if microbial life had taken root, then there was a chance it might still be living in the soil of Mars today. The astronomer Carl Sagan promoted optimism in continuing the search for life on Mars, and even speculated that visible forms of life might still be found at or beneath the surface.[42] The question of life would be the primary objective of the upcoming Viking lander mission. Even many of the scientists who were skeptical about the search for life still saw value in learning what Mars geology and chemistry could tell them about the evolution of Earth as a habitable world.[43]

### VIKING SEARCHES FOR LIFE

There are several interesting counterfactual "what if" questions one can ask about the Cold War history of spaceflight, and about how small contingencies might have changed the course of the space race. What if the German army hadn't gambled on the power of the rocket? What if the United States had placed a satellite in orbit before Sputnik? And what if Viking had determined unambiguously that life existed on Mars? This last question has stimulated speculation that human missions to Mars would have followed—that Paine would have gotten his wish of seeing a new Apollo for the 1980s.[44] But as it happened, the Viking mission failed to settle outstanding questions about the presence or history of life on Mars. It would be the last Mars mission of this first era of robotic exploration.

It would be a mistake to blame Viking for temporarily taking the wind out of the sails of Mars exploration. No mission since Viking has settled the life question, and yet Mars exploration is alive and well today. As we've seen, the decision to scale back NASA ambitions was made years earlier by the Nixon administration. The United States did not stop exploring Mars because Viking failed to live up to expectations. It paused its exploration of Mars because big-budget planetary exploration had, at least temporarily, mostly exhausted its political capital. It's possible that the discovery of life could have put new money in the bank, but this is far from a certainty.

In many ways, the Viking mission was a great success. Technologically, the two landers were far more complex than NASA's previous Surveyor lunar soft landers. Unlike a Moon lander, a Mars spacecraft has to work independently and autonomously: it takes anywhere from four to twenty-four minutes, depending on where Earth and Mars are in their orbits, for a signal traveling at the speed of light to make the one-way trip between Earth and Mars. Landing safely becomes one of the greatest challenges—one that terrifies spacecraft engineers to this day. Fortunately, computers had advanced a great deal since the Surveyors were designed in the 1960s. By the early 1970s the development of the integrated circuit (commonly known as the microchip) had drastically reduced the size of computers while increasing their capabilities. Engineers could now design a spacecraft that could run a large number of operations without much human intervention, provided it was properly programmed.[45]

In the summer of 1975, four NASA spacecraft flew to Mars on a ten-month trip; two life-seeking Viking landers were sealed inside protective "bioshield" shells and were attached to two Viking orbiters. In addition to protecting the lander during the journey, the bioshield was meant to protect the sterilized lander from contamination by bacteria on Earth before launch. This measure was deemed necessary to protect the planet from possible microbial contamination and, more importantly, to prevent potential false positive results for the onboard life-detection experiments.

Once at Mars, the Viking orbiter image teams looked for safe landing sites for the landers, more closely examining candidate sites selected on the basis of Mariner 9 maps—sites that had appeared to be free from hazards. This turned out to be a more difficult task than first imagined. The Viking orbiters carried better cameras than Mariner 9. When they examined the candidate landing sites, they found that the sites were not as safe as they'd appeared; they were strewn with large boulders. The orbiters spent a month looking for completely new landing sites.

Once new sites were selected, the most nerve-wracking part of the mission began. Each lander had to make its way from orbit to the surface. One part of this journey was easy: falling. Once the lander was detached from the orbiter, rockets slowed its orbital velocity so that gravity could pull it toward the ground in its bioshield. The bottom of the bioshield protected the lander from the heat created when, traveling thousands of miles per hour, it entered the Martian atmosphere. Next, the lander had to slow down even more. A parachute would do some of this work, but the Martian atmosphere is thin and a parachute can only do so much. When the lander was about a mile above its landing site, it used a set of rockets to slow its descent enough for a soft landing. The two landings went flawlessly; years of development and testing paid off. The computer-controlled entry, descent, and landing sequence—the template for similar landing sequences used on every Mars lander and rover that followed—proved itself for the first time (figure 12).

The primary experiment package on Viking was designed to search for life. Viking's chief project scientist, Jerry Soffen, was very interested in exobiology (the search for life beyond Earth) and in developing tests for life on other planets. In his view, the real test would be whether organic molecules could be detected on the planet's surface. Life as we know it requires organic molecules and produces organic molecules—lots and lots of organic molecules. On Earth these organic molecules become part of a circular food chain that extends from microbes to plants, to animals, and back again. Soffen was adamant that if life existed on

**FIGURE 12** Viking carried a US flag to the surface of Mars, mounted on the housing of its nuclear power system. The lander took this image of itself on Mars, showing the flag. NASA.

Mars, the best way to detect it would be to test for organic molecules and the metabolic processes that produce them. "That's the ball game," he said. "No organics on Mars, no life on Mars."[46] Each lander's robotic arm would dig trenches in the Martian dirt and deliver samples to experiments designed to stimulate and detect microbial metabolism.

The first of the three life-detecting experiments was the gas exchange experiment (GEX). In a chamber filled with inert helium gas, the GEX added a nutrient broth to the Martian soil. The experiment sampled the atmosphere periodically to determine whether any new molecules had been released by any organism eating and digesting the nutrients. The experiment assumed that, like organisms on Earth, a Martian microbe would release oxygen, carbon dioxide, nitrogen, methane, or some other organic gas in the course of metabolizing its food.

The second experiment was the labeled release experiment (LR). Like the GEX, the LR chamber added a nutrient broth to the Martian soil and ran tests to see whether anything was happening. All of the nutrients used in this experiment were organic molecules tagged with a radioactive isotope of carbon. The air in the chamber was monitored to see whether any radioactive carbon dioxide was produced by a microbe eating and metabolizing the nutrients. This was based on a common test used by sanitary engineers on Earth to test water samples (though not usually with radioactive carbon), and the Viking version of this experiment was designed by a sanitary engineer named Gilbert Levin.

The third experiment—the pyrolytic release experiment (PR)—was designed to test whether there were any photosynthetic microbes in the soil. Like the LR experiment, this test used radioactive carbon—not to label the nutrients in a broth, but to label carbon dioxide and carbon monoxide gas introduced into the sample chamber. The experiment gave the Martian soil water and light, and waited to see whether the carbon in the chamber's atmosphere was incorporated into the growth of anything living in the sample. On Earth, plants and photosynthetic bacteria would take up the carbon from the gases in the course of making energy out of sunlight; the experimenters assumed that a photosynthetic microbe on Mars would do the same. At the end of several days of incubation, the gases were removed from the chamber, the soil was "baked" at a high temperature, and the radioactivity of the soil was measured to see how much carbon, if any, was now in the soil.

Two of these three experiments, the GEX and the PR, returned negative results. They detected no evidence that anything was living in the soil they tested. The LR, however, seemed initially to produce a positive result. After the nutrient broth was introduced to the sample for the first time, radioactive carbon was detected in the atmosphere of the chamber. This at first made the exobiologists optimistic. Maybe there was something alive in the soil. However, when a second dose of the broth was injected into the chamber, no additional radioactive carbon was detected in the atmosphere. Had something really been living in the soil sample, it would almost certainly have continued to eat and metabolize the labeled nutrients and produce the same gases. The consensus was that the initial detection of life had been a false positive caused by some chemical property of the Martian soil. Gil Levin was not convinced; he remained adamant that his experiment had detected life at both landing sites.

The life-detection team was only one of three science teams attached to the landers. Viking was also designed to detect and measure the minerals that composed the dirt and rocks at its landing site. Each lander carried an experiment designed to analyze organic chemistry (the carbon-based chemistry associated with life on Earth), as well as one geared toward inorganic chemistry (everything else). The organic chemistry team's experiment was a gas chromatograph–mass spectrometer (GCMS). Like the biology experiment, the GCMS relied on the lander's robotic arm to deliver a sample of Martian soil to a chute on the top of the lander. Once the sample was inside the GCMS, it was baked in a tubular oven to release any gases present. These gases were separated using a gas chromatograph, and their composition was analyzed with a mass spectrometer (an instrument that separates molecules and atoms based on their atomic weight). One significant finding of the GCMS was that no organic molecules were detected in the soil samples it analyzed at either landing site. The instrument's principal investigator suggested that the gases detected in the LR experiment must have been caused by something in the soil that had nothing to do with life.

The third science team attached to the Viking landers used an X-ray fluorescence experiment to study the inorganic chemistry of Mars. This instrument bombarded soil samples with X-rays and then measured the angles at which the X-rays bounced off the grains of dirt. Significantly, the inorganic chemistry team determined that Mars was covered in an iron-rich clay that also included silicon and aluminum. This, combined with Mariner 9's discovery of channels, gave many geologists hope that Mars had once been wet, as clays on Earth are typically weathering products that require water. While the biological and organic experiments indicated no signs of life, the mineralogy of Mars spoke to past habitability.

**PLATE 1** The "Mars Beast" as it appears in the Dresden Codex. SLUB.

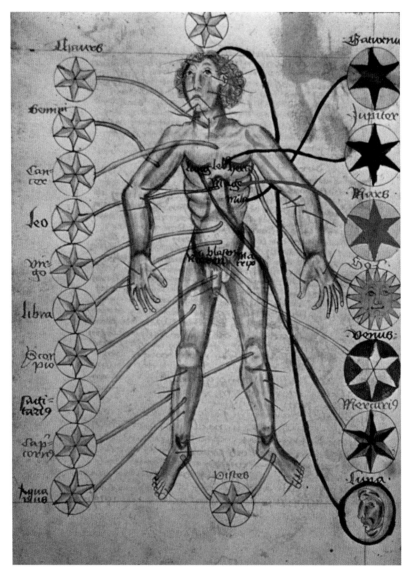

PLATE 2  A zodiac man, from a fifteenth-century German manuscript. MacKinney Collection of Medieval Medical Illustrations.

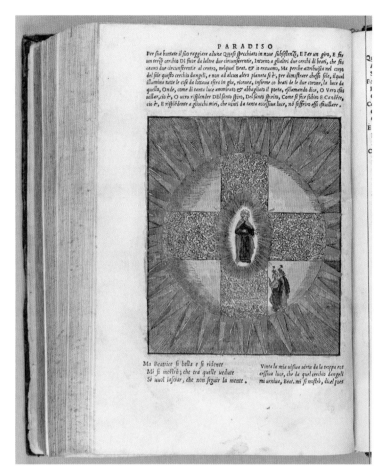

**PLATE 3** A sixteenth-century illustration of Dante's visit to Mars. B85DL B44, Rare Book and Manuscript Library, Columbia University, New York.

**PLATE 4**  Henrique Alvim Corrêa's illustrations in a 1906 large-format French edition gave life to the Martians described by H. G. Wells in *The War of the Worlds*.

**PLATE 5** Eric Brevig and Alex Funke in front of the miniature sets built as backdrops for the 1990 film *Total Recall*. Image courtesy of Eric Brevig.

PLATE 6 The Sojourner rover examines a rock on Mars. NASA/JPL.

**PLATE 7** Artist's concept of the Mars Exploration Rover on Mars. NASA/JPL.

**PLATE 8** Three generations of Mars rover in the JPL Mars yard. Sojourner's flight spare, Marie Curie (in foreground), with testbed versions of the Mars Exploration Rover (left) and the Mars Science Laboratory (right). NASA/JPL.

**PLATE 9** The Perseverance rover's parachute, seen from an "up-look" camera on the rover's back shell during landing approach. Two messages printed on the chute in binary code give the coordinates for the Jet Propulsion Laboratory in southern California, and the JPL motto: "Dare mighty things." NASA/JPL.

**PLATE 10** Artist's concept of the Perseverance rover and the Ingenuity rotorcraft on Mars. NASA/JPL.

**PLATE 11** NASA artist's concept of the first humans on Mars, including a modular habitat and a roving vehicle. NASA.

**PLATE 12** The desolate and inhospitable surface of Mars, as witnessed by the Perseverance rover. Although it superficially resembles the American Southwest, there are none of the comforts of home on Mars. NASA/JPL.

## MARS: A WORLD OF ROCKS

The two Viking orbiters contributed to the evolving understanding of Mars's geologic past initiated by Mariner 9. Though the orbiters were not intended to map the entire planet, they survived long enough to accomplish the task. They spent almost two full Martian years (close to four Earth years) observing Mars. The Viking orbiter camera system, along with the infrared thermal mapper (IRTM) experiment, monitored weather patterns, global dust storms, the redistribution of fine-grained materials on the surface, and the seasonal advance and retreat of the frozen polar caps. The IRTM was designed to allow researchers to determine the rockiness of different regions during landing site selection, but it could also be used to distinguish carbon dioxide frosts and ices from those of water. Orbiter photographs combined with IRTM data refined geologists' understanding of the polar caps, showing that they changed significantly with the seasons. Geologists also observed large-scale weather patterns in the fluctuating Martian temperatures, similar to continental weather patterns on Earth. Most interestingly, the IRTM data from the polar caps suggested that during the Martian summer, when the carbon dioxide ice retreats, the ice that remains contains frozen water. If Mariner 9 discovered that Mars was a planet with its own geologic history, Viking conducted the first meaningful exploration of this new world.

Planetary geologists in the post-Viking years continued to explore Mars through the vast amounts of images and infrared data collected by Mariner 9 and Viking. The geologists pressed on, even as no consensus emerged from NASA and its advocates about what Mars missions should come next.[47] They mapped the stratigraphy of Mars for the first time using a combination of Mariner 9 and Viking images. On the basis of this stratigraphy and careful counts of craters in different regions, they gave Mars its first geologic eras (figure 13). They broke Mars's geologic history into three main periods—Noachian, Hesperian, and Amazonian—which they went on to divide and characterize further over the next two decades.[48]

**FIGURE 13** Geologists at the US Geological Survey produced the first detailed geologic map of Mars using Mariner 9 imagery, depicting its geological ages for the first time. USGS.

In 1981 the geologist Michael Carr, who had served as a member of the Mariner 9 imaging team and as leader of the Viking orbiter imaging team, wrote what would be perhaps the most influential work on Martian geology for the next fifteen years: *The Surface of Mars*. In it he attempted to summarize the surface geology of Mars as revealed over more than a decade of spacecraft observation. The first half-billion years of Mars's geologic history had been erased and reshaped, most likely by large impacts. After the crust solidified, the rate of impact had tailed off and the landscape had become stable. The oldest parts of Mars were in its southern hemisphere, where the cratering record was most intense. A younger surface laid down by volcanism was found in the northern hemisphere. What exactly had caused the differentiation of the two hemispheres was not yet understood.

At the time of differentiation, the volcanoes of Tharsis were probably already active, but perhaps were still growing into the massive volcanoes they would eventually become. The runoff channels were very old, perhaps as old as 3.9 billion years, with numerous tributary systems in the oldest regions of Mars. But they disappeared in terrain estimated to be less than 3.5 billion years old. Volcanism had likely continued on Mars for a very long time, bringing iron-rich materials to the surface, but for the past billion years it had been limited to the Tharsis region.[49] Carr took this to imply that Mars may have had a thicker and warmer atmosphere only in the earliest period of its history, and that it hadn't lasted long. Where the atmosphere went was unknown; maybe it was bound up in the rocks or in the ice, as permafrost or subsurface glaciers. Carr believed it likely that water remained in "a vast artesian system" below a thick layer of permafrost, trapped there ever since the atmosphere had thinned. As for the outflow channels, he speculated that something might have caused periodic breakouts of this trapped water. However, the water could not have seeped back into the ground because of the permafrost layer, and so the question of where the water had gone was unanswered. Maybe there was ice, mixed with volcanic debris, in the high latitudes.[50]

Carr and others would spend the time between Viking and the missions of the mid-1990s picking apart the similarities and differences between Earth and Mars. The surfaces of both planets showed evidence of modification by volcanism, wind and water erosion, and chemical reaction with the atmosphere. And yet the surfaces told very different stories. In some ways, the story of Mars did help scientists understand the history of the solar system, by revealing periods of formation and bombardment obscured by Earth's more dynamic surface and plate tectonics. But the dynamics of the Martian crust were completely different—showing no evidence of lithospheric recycling, or of Earthlike plate tectonics. Mars was "an active planet with enormous surface relief, on which features with a wide range of ages are preserved in almost pristine condition."[51]

In their textbook *Earthlike Planets*, Bruce Murray, Michael Malin, and Ronald Greeley presented their argument for a comparative planetology of the inner solar system based on "a common planetary environmental history." They catalogued the similarities and differences between Earth, the Moon, Mercury, Venus, and Mars. From this they derived a general history of the formation and evolution of the rocky planets which incorporated evidence from Apollo lunar samples, spacecraft photography and data, and insights gained on the surfaces of the Moon and Mars.[52] The picture was continuous, but still full of contradictions and questions. Their chapter "Comparative Planetology" looked to the future for more answers. They imagined mobile robotic systems that could be sent to explore Mars's surface, which "might be designed to mimic to some extent the role of a field geologist making traverses over the countryside."[53] If such a thing could happen, with Earth-based geologists providing the brains for the robot as it sent back TV signals to Earth, they envisioned "a golden age of geographical and geological exploration of the surface of Mars."[54] They also imagined that, given enough time—and they admitted it could take generations—humans would eventually set foot on Mars to begin an era of human exploration.

### MARS IN THE POPULAR IMAGINATION

The late 1940s and 1950s saw an explosion of depictions of the human future in space in magazines and on television, and these increasingly argued that the interplanetary human would soon go from science fiction fantasy to reality. They more than implied that for humans to become true spacefarers would be not only a great technoscientific accomplishment but a necessary step in human cultural progress, or even human evolution.[55] As mentioned above, figures like von Braun teamed up with artists and popular writers to create technologically sophisticated—if perhaps not entirely feasible—visions of future Mars missions. The "visioneers" who promoted even more far-futured design and engineering concepts promised that peace and abundance would be realized in (or through) space after lunar and interplanetary spaceflight had been achieved.[56] The idea of Mars as a human world became an important piece of these new futures.

Space-based science fiction programming took off in the 1950s and '60s. Whether informational or entertaining, optimistic or bleak (one fictional TV family found itself *Lost in Space*), these shows presented space as the future not just for astronauts but for domestic life. They also tended to be utopian visions of a unified human species. Cold War tensions were translated into conflicts or standoffs between humans and aliens, but these future humans had mostly moved past national or cultural rivalries. In the children's program *Space Patrol*, this ideal of unity is literally built into the artificial planet on which the United Planets governing body is headquartered, the core of which is composed of soil from the home worlds of member species. In the 1960s, *Star Trek* continued this theme; alien contact led not to the colonization of Earth by superior aliens, but to the welcoming of an immature but promising human species into a United Federation of Planets. The Star Fleet presented in *Star Trek* was military in organization, but its mission was ostensibly peaceful; and between episodic exploits, life aboard the *Enterprise* was similar to life in the large organizations that typified mid-twentieth-century American life. Notably, the show presented a view of the future in which humans, not robots, performed nearly all work in space.

No longer confined to science fiction and fantasy, spaceflight found its place in Cold War science television programming. If von Braun and Disney's *Man in Space* collaboration initiated this first era of space on television, Carl Sagan's *Cosmos* series was its most memorable successor. Over the course of the Mariner and Viking missions, Sagan emerged as one of the most outspoken advocates of planetary exploration and its value to humanity. He could often be seen on late-night talk shows, where he presented a charismatic face for space science. In 1980, after the end of Viking and the first Pioneer and Voyager encounters with Jupiter, he appeared as host of *Cosmos*, a television series that brought viewers along on a journey through time and space to show them their place in the universe. Sagan's attempts to make space science and exploration engaging also included popular books and even speculative science fiction novels that explored the human relationship to space. These works combined a romantic view of exploration with a cautiously hopeful message that scientific discovery and understanding could overcome the Cold War militarism that had initiated spaceflight. In the book that accompanied *Cosmos*, Sagan warned that we should not export our destructive tendencies to Mars. "What shall we do with Mars?" he asked. "There are so many examples of human misuse of the Earth that even phrasing this question chills me."[57]

Even with the proliferation of television and film, print literature still carried most of the weight when it came to speculation about the meaning of spaceflight and its connection to human history. The genre of science fiction, having developed from diverse strands throughout the nineteenth and early twentieth centuries, had coalesced in the 1920s and '30s and was now a thriving industry. Many of these authors bucked the popular trend of connecting the future of Mars exploration to a romanticized history of geographic discovery and frontier exploration. In his pre-Mariner novel *The Sands of Mars* (1951), for example, Arthur C. Clarke presents a Mars with two possible futures. The native Martian life can be preserved and allowed to follow a natural evolutionary path, or humans can drastically reshape the planet through a grand terraforming project. Like Sagan, Clarke was preoccupied with the question of transcendence: Will moral advances accompany the technologies that make spaceflight possible and allow us to escape our past, or will the future be characterized by the same opportunism and exploitation that defined European and American expansion?[58] Clarke seems to have hoped his readers would consider the actual violent history of settler colonialism, not just romantic or patriotic narratives of pious pilgrims and pioneering frontiersmen.

One of the most influential pieces of Mars fiction during the pre-Mariner era was Ray Bradbury's *The Martian Chronicles* (1950). If Clarke's novel can be read as being hopeful that the space age would eventually escape its military origins, Bradbury's work seems to speak more to the pessimistic view. The loosely related collection of stories that made up the *Chronicles* presented humans in the American mold who travel to Mars not to escape history, but to recreate it. Bradbury's Mars is not a utopia; it is a landscape on which to reenact and de-romanticize the failings of previous periods of expansion and colonization. It also becomes a vantage point from which its new inhabitants can watch their home world being destroyed by global nuclear war. This is a book that reportedly inspired many in the planetary science community, including some of my colleagues, to want to explore Mars. They nostalgically remember reading these stories through youthful eyes that didn't understand them as colonization allegories, but which instead fixated on the idea of experiencing another world.[59]

These pre-Mariner versions of Mars presented the planet as only barely habitable. Like their peers in the scientific community, science fiction authors believed the planet had an atmosphere perhaps one-tenth the density of Earth's. Life there would be challenging, like living at the top of a high mountain range, and would require breathing masks, warm clothes, and pressurized living spaces. Martian life would be difficult, but it could be endured. The revelations of Mariner shattered this illusion. With a carbon dioxide-rich atmosphere less than one percent as dense as Earth's, the planet was unlivable without drastic intervention. Either it would have to be terraformed into an Earthlike world, or life on the red planet would be limited to special habitats or domed cities. Robotic exploration of Mars put some severe constraints on what journeys to Mars were imagined in the late twentieth century.

But why send humans to Mars in the first place? Scientific exploration was one answer, but increasingly capable robots were already doing that. When alternative motivations are considered, aspects of the Cold War separate from missile and surveillance technologies come into play. Cold War America valorized capitalism and attempted to spread the reach of its markets to as many nations as it could. Could Mars be a place for capitalist intervention? Science fiction authors believed that it could. Very early on, writers imagined corporate-owned Mars colonies. In his 1949 young adult novel *Red Planet*, for example, Robert Heinlein imagined Mars in the vein of a company town in which all aspects of life were governed by an Earth-based interest. By the end of the twentieth century, this image of Mars as a place where humans live their lives in the service of an absentee landlord became even more common, as visions of space and capitalism grew more pessimistic. At the movies, both space and the future had become distinctly less utopian. Late-twentieth-century movies like *Alien* (1979) and *Blade Runner* (1982) depicted spaceflight as a profit-seeking venture willing to sacrifice human lives, or as a means of escaping an overpopulated, polluted, and resource-stripped Earth.

One of the first blockbuster film depictions of Mars followed this bleak trend, extending what one scholar identified as science fiction cinema's "critique of the multinational state and corporation" to the red planet.[60] The 1990 science fiction film *Total Recall* was based on the 1966 Philip K. Dick short story "We Can Remember It for You Wholesale." Dick's writing is concerned with the exploitative force and violence of capitalism on Earth, and he presents Mars as an elite destination most humans simply can't afford to visit. The main character, Douglas Quail, lives in a futurist Chicago defined by hovercars and robot cab drivers. He dreams of Mars, yet can only barely afford "ersatz planetary travel"— the implanting of "extra-factual memories" of having vacationed on other worlds. The movie *Total Recall* took Dick's premise of fractured Mars dreams, memory, and reality and ran with it. The director Paul Verhoeven had already established himself as a master of consumer-capitalist techno-dystopia with the 1987 film *RoboCop*. With a script originally written by the team that wrote *Alien*, Verhoeven's Mars was not just out of reach to the average Earthling; it was an exploitative mining colony that starved its impoverished workers (Earth's dispossessed) of oxygen and protection against radiation. The oppressed masses of Mars threaten to rise up, but are kept at bay through these deprivations.

Dick's story was written before Mariner, and it populated Mars with simple lifeforms like cacti and "maw worms." It was "a world of dust where little happened, where a good part of the day was spent checking and rechecking one's portable oxygen source."[61] The film diverged from the original story to accommodate the Mariner and Viking revelations that Mars was not just arid but lifeless. The Mars we see in this film feels strikingly real, even if the red hue of the planet is oversaturated and the action sequences are over the top and intentionally campy. The film presents an elaborately designed vision of what a Mars colony built around a commercial mining operation might look like.

Our first real glimpse of the planet is as Quaid (no longer Quail, as in Dick's original story) arrives on Mars and leaves the Martian spaceport on a ground transport train. The Mars we see in these scenes is in fact a composite of a warehouseful of miniature sets ranging from one-twelfth to one-two-hundredth scale, made over the course of a year by a large team of Hollywood model makers (plate 5). The natural features of Mars were built from chicken wire and surfboard foam, covered in old carpet, and dressed with sand, stucco tint, and red mortar dye. Detailed miniature structures populate the sets—a Hilton hotel and a skyscraper office building provide familiar landmarks, trains and elevators allow movement across the surface and in and out of vast canyons, and mining equipment is visibly at work fulfilling the colony's commercial purpose. This Mars is glimpsed throughout the movie, mainly through the windows of the habitable structures where the action takes place.[62] It was one of the first movies to present Mars in this way while simultaneously being one of the last to use matte painting and miniature sets rather than CGI computer graphics. In this regard, the film outshines many of the movies that came after it for its hand-crafted attention to gritty detail in the construction of the sets and backdrops.

The Mars of *Total Recall* feels, at least visually, as though it might occupy the same future as other dark science fiction films of the era. This is no coincidence. The film's conceptual artist, Ron Cobb, who designed all the technology seen in the film, from spaceships to full-body X-ray security scanners, had previously served in the same role on *Alien* and its sequel *Aliens*, as well as on other well-known science fiction movies. Miniature effects cosupervisor Mark Stetson had likewise previously worked as the chief model maker on *Blade Runner*, where he met *Recall*'s visual effects supervisor, Eric Brevig (then an intern).[63] Working across a series of unconnected science fiction films throughout the 1980s, these artists presented a somewhat coherent vision of the tech that would define the future. But *Total Recall* also lives in the real world of Mars exploration. The artists who worked on the film felt that Stanley Kubrick and Arthur C. Clarke's *2001: A Space Odyssey* (1968) had "raised the bar" in presenting a realistic—some might say prophetic—view of the future based on "the science of the day and the knowledge of what future technologies were being researched." Cobb was one of a handful of artists who had become known for having "the knack of blending imagination with science to design fantastic or future technologies."[64]

Cobb and the miniature artists brought with them their own experiences of Mars. They had lived through the Mariner and Viking era of exploration, and, working in Hollywood, were not far from the center of planetary exploration, JPL's Pasadena campus. Shortly before preproduction on *Total Recall* began, Cobb contributed to an unfinished Mars film project initiated by Apple Inc., where he met JPL scientist and Viking alumnus David Pieri, who was acting as a pro-bono consultant.[65] Pieri and the team from Apple brainstormed a fictionalized narrative based on human Mars mission concepts while Cobb sketched concept art of the required technologies. Cobb most likely carried some of what he learned to *Total Recall*.

Stetson's father occasionally worked with JPL, and he remembers seeing the Viking lander surrounded by a photo exhibition of Viking images during a 1970s visit to the lab. This gave him a lasting impression of the reality of Mars, which he imparted to his miniature sets. Stetson was also influenced by childhood visits to the 1964 World's Fair in Queens, New York, and Expo 67 in Montreal, Canada, where he became fascinated by the avant-garde architecture of the conceptual space settlements he saw there.[66] The mining equipment Cobb designed for *Total Recall* was inspired by similar equipment on Earth, and Stetson and his team produced the miniatures using the "kit bashing" tradition of stealing parts from existing models of industrial machinery and combining them in new ways on custom builds. These familiar technologies were reimagined by a team who had grown up consuming and producing images of future space technologies.

Robotic exploration provided one new twist that the filmmakers exploited. One Viking orbiter image had caught the public's attention and raised speculation not only about past Martian life, but about ancient alien civilization: a rock formation in the Cydonia region seemed to have the shape of a recognizable but not-quite-human face (figure 14). Scientists provided explanations for how the face and other seemingly artificial structures were in fact the products of sunlight hitting natural rock massifs and casting suggestive shadows. Still, many in the public persisted in believing that an alien civilization—either native to Mars

**FIGURE 14** Viking Orbiter image of the facelike rock formation it discovered on Mars.
NASA/JPL.

or transplanted from another world—might have left evidence of its existence on the red planet, possibly even intending for humans to find it.[67] *Total Recall* played with this public fascination concerning a vanished civilization on Mars. In the film, the face on Mars is replaced by a mysterious ancient alien technology built into the interior of a dead volcano—revealed to be a reactor capable of melting the planet's icy core and making it a habitable world. When Quaid activates it, it rapidly gives the planet an Earthlike atmosphere and temporarily liberates the workers from their capitalist overlords.

In Dick's original story, a Mars memory implantation kicks off the story's second act, as it triggers Quail's real memories that he in fact has been to Mars, as an "Interplan" assassin sent to kill a high-ranking official. Our mild-mannered protagonist has experienced Mars as a highly trained tool of the state. The film, more than Dick's original story, played with the idea that even after this revelation, Quaid's trip to Mars may yet be completely imaginary. The reawakened Quaid travels to Mars, believing that he is fighting for the Mars resistance, only to find that his past self was in league with the corporation and misled his new self so that he could bring down the resistance from within. Quaid is presented with the option to play hero or villain. But even this may be an implanted fantasy. Was Quaid a man with a dual personality, choosing to heroically sacrifice his real identity to save Mars, or was he a consumer of Mars memories who had paid to experience an adventure? Verhoeven's film provides no answer, questioning identity and reality, and ultimately abandoning both for an adventure with a dramatic ending.

### MARS AFTER DETENTE

In 1991, the year after *Total Recall* was released, the Cold War ended. Before the end of the decade, the twenty-year hiatus in robotic exploration of Mars would also end. A new era of increasingly capable orbiters, landers, and rovers would explore the different regions of Mars. A veritable fleet of robots would be sent to the red planet. The dream of operating mobile science platforms and traversing the ground, conceived in the Cold War, would be realized in an entirely different context. The success of this dream would rely not on geopolitical competition, but on public support and engagement. People—not just scientists—would come to know Mars and its robotic explorers as never before.

As we will see in the next chapter, the new science of Mars would be dramatic. New methods in the search for life or evidence of past life on Mars would make the Viking experiments look technologically sophisticated but conceptually premature. Just as Mariner 9 and Viking had proved the image returns of the previous Mariner missions to have been misleading, new experiments done in the Martian soil, combined with studies of extreme environments and microbes on Earth, would call into question Viking's assumption that extant life might be sitting on the surface of Mars.

# *MARS AND THE NEW MILLENNIUM*

**THE HOME WORLD MOURNS A ROVER**

On February 12, 2019, a team of engineers at JPL sent one final set of commands to the Mars Exploration Rover (MER) Opportunity. Affectionately nicknamed "Oppy," the rover had been exploring Mars for fifteen years (figure 15). It had proven itself to be a resilient robotic adventurer, surviving broken wheels, sand traps, and the harsh Martian climate. But now it had been unresponsive for months. Its final message to Earth arrived on June 10, 2018, on the eve of a giant dust storm that darkened the Martian sky. The storm, which Oppy had first detected at the end of May, grew dense enough to cut off sunlight to the rover's solar panels. The mission team hoped that Oppy could ride out the storm in hibernation mode, using minimal power to keep itself warm. But the final downlink indicated that the rover's battery was close to dead. To those who worked with her, Oppy wasn't an "it" but a "she," and they knew that without battery power, she would have no way of protecting her aging circuits from the cold.

The final transmission from the rover noted that the high levels of dust in the atmosphere had turned the sky dark. The dust storm lasted for two months, and during this time JPL received none of the automated pings from Oppy that mission controllers were expecting. They

**FIGURE 15** Opportunity's first "selfie" on Mars. NASA/JPL.

didn't give up all hope, but they knew that her battery had most likely died. They watched as robotic spacecraft in orbit around Mars tracked the storm, waiting to see the surface features of the planet emerge from the cloud of dust. As the Martian sky finally cleared toward the end of June, JPL engineers expected that it could take months for Oppy's solar panels to recharge her. But one final problem presented itself: Oppy's solar panels were blanketed with dust, unable to generate sufficient power to bring the rover back online. Though they tried for eight months to revive her, the mission controllers eventually had to let her go.

When the news spread that NASA was giving up on reaching Oppy, the science writer Jacob Margolis translated her last bit of data (with some poetic license) into the message, "My battery is low and it's getting dark," and posted it on Twitter. Oppy's "last words" were quickly made into memes and cartoons that were tweeted and posted widely on the Internet. Along with these came heartfelt thank-yous for the many years of exploration and discovery. The well-wishings came not just from those who had designed, built, or operated the rovers, but from people who had enjoyed hearing about the discoveries made on Mars. For some who had grown up with Oppy on Mars, losing the rover was like losing a piece of their childhood. Margolis remembered first being shown the rover's images as a kid: "I was enraptured by the promise of NASA's most ambitious rover mission yet and that we could potentially confirm that water, and maybe even life, once existed there. It's one of my favorite science memories."[1]

Many of Oppy's mourners perhaps knew very little of the esoteric knowledge the rover helped collect, but nonetheless were impressed by Oppy's seeming tenacity, were inspired by the dramatic horizons captured by her panoramic imaging system, or had enjoyed her occasional "selfies" atop rocky outcrops. As NASA had learned over the previous quarter century, the public engages with rover missions in ways not inspired by flyby or orbiter missions, and the desert-like landscapes of Mars, even if devoid of alien civilizations, can be used to evoke romantic notions of exploration and discovery.

Oppy's fans most likely interacted with her through channels that NASA neither controlled nor had anticipated. Spirit and Opportunity had a web presence, as had their predecessor rover, Sojourner. In the era of Mariner and Viking, Mars coverage had appeared sporadically in newspapers, television, and radio—mostly limited to launches, arrivals, and occasional press conferences. But home computers and later smart devices allowed the always connected public to participate in the rover missions, providing immediate access to returned images and news about what the rovers were doing. These platforms tended to emphasize the rovers and their exploits on Mars, not the teams working on Earth to operate them, or the teamwork required to decide what they should do from day to day. The rovers seemed to speak for themselves.

NASA began experimenting with Twitter in 2008, when the JPL public relations office set up its first account to report the activities of NASA's Phoenix Mars lander. To stay within the short character limit of the new platform, Veronica McGregor, who managed the account,

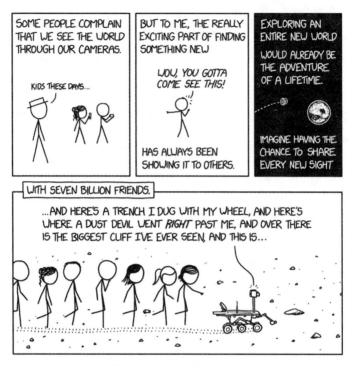

FIGURE 16 "Opportunity Rover" web comic, XKCD.com.

chose to write the tweets in the first person, beginning each one with
"I" instead of "The spacecraft." She gave the lander a voice, and created
what could be read as a direct line from Mars.[2] Even without her own
Twitter account, Oppy was operational long enough that she earned
her place in popular culture. A satirical Twitter account in the name
of @SarcasticRover (maintained by the screenwriter Jason Filiatrault,
and written in the persona of a salty counterpart to Spirit and Opportunity) actually managed to attract a couple thousand followers during
the rover mission. The two rovers also appeared regularly in web comics
like Randall Munroe's *XKCD*. In one comic, written after Oppy's eleventh
year roving the red planet, the rover was imagined as an unstoppable
force with which humans would one day have to share Mars. In an earlier comic, she had also appeared to be no different from a human with
a smartphone, sharing her experiences of Mars with her followers on
Earth (figure 16).

After NASA declared Oppy dead and the mission over, her internet followers began to mourn her loss. @SarcasticRover tweeted, "It's just a fond farewell to a friend. From a pretend rover to the best rover." In the *Washington Post*, Filiatrault wrote, "I hope somewhere in your diodes and processors, mixed in with all your geological sensors, you could sense the inspiration you gave humanity. I hope you felt the love they had for you—the way you made science seem so close and so possible. The way you made a planet millions of miles away seem as close as next door."[3]

One of my favorite responses to the news that Oppy had died was the recirculation of an altered *XKCD* from 2010, when Opportunity's twin, Spirit, had gotten stuck in the sand and died. The ten frames of the original version tell the story of Spirit's success on Mars, outliving her original mission time frame and continuing to explore Mars (figure 17). At the end of her journey, unable to rove any farther, Spirit asks if she can return home. After all, hasn't she done a good job? Wasn't she a good rover? Perhaps because the JPL mission controllers don't know how to

**FIGURE 17** "Spirit," XKCD.com.

break the news to her that she will never see Earth again, no response is returned. In the altered version, XKCD fan artist "Burkitt" added two new frames to the end of the comic, in which many more years pass. Eventually a human explorer in a recognizable spacesuit arrives to visit Spirit. Even more years pass, and the rover now sits on a small circle of Martian soil, the area around her built up into a habitable Mars colony enclosed in what looks to be a transparent geodesic dome.[4] When Oppy died, the fans returned to the threads in which they'd originally mourned Spirit. They rehashed conversations they'd had nearly a decade earlier, and revisited—perhaps reexperienced—their surprise at feeling such empathy for a piece of machinery.

### THE CONTEXT OF POST-COLD WAR MARS EXPLORATION

This extended discussion of the public reaction to the death of Opportunity illustrates the main themes of this final chapter in the history of Mars exploration. Since 1997, because of longer-lived spacecraft and a mostly continuous stream of missions, Mars exploration has been a largely uninterrupted activity introducing a new cast of orbiters, landers, and rovers. This enterprise has produced more data, in more spectral wavelengths and at higher resolution, than was achieved in the first decades of exploration. It has transformed Mars into a "known world"—one that still holds questions, to be sure. The orbiting missions have done the lion's share of the science, at least when it comes to geographic coverage and description of planetary-scale phenomena. But it's the landers, and especially the rovers, that have engaged the public imagination. The fact that this exploration has been done robotically, and without human "boots on the ground," has not, by and large, dampened the public's appetite for exploration—an appetite fed by new media that provide constant access to images and stories, and which encourages follower vicariously to imagine themselves in the Martian landscape.

The robots themselves are often presented as intrepid explorers, reminiscent of an earlier era of terrestrial exploration and the romantic notions that went along with it—one in which human explorers risked their lives to reach the inhospitably extreme environments of the North and South Poles. Oppy's "last words" recall those of the seasoned adventurer Lawrence Oates, who died on Robert Scott's *Terra Nova* expedition to the South Pole (1910–13). Suffering from frostbite and gangrene, Oates knew that his lingering on was only putting the other members of the expedition at risk. He walked out of his tent into a blizzard, telling his crew, "I am just going outside and may be some time." Scott interpreted this act of self-sacrifice as the ultimate example of a British army officer's bravery and resolve.[5] Or consider another arctic explorer, Alfred Wegener—famous today as the originator of the theory of continental drift—who died in Greenland in 1930 while attempting to resupply his expedition. He had insisted that exploratory science demanded heroics. Clues to the inner workings of the universe were, he believed, much more significant than himself.[6] His men built an ice-block mausoleum around him, with a large iron cross to mark the grave. Like Oates and Wegener, Oppy now rests in the spot where her exploration ended.

Rovers and their imaging systems have made Mars seem closer than ever before, and have allowed even greater fidelity in our imaginings of humans on Mars. The public mourning of Oppy speaks to an expectation that robotic exploration of Mars is connected to the human future—to human exploration of Mars, or even human settlement. This would be the culmination of some of the earliest musings about human expansion to other worlds, and to the human ability to bend even alien nature to our will. Since Mars is now known to be incredibly inhospitable to life—at least the animal and plant life on which humans depend—such an endeavor would likely be the most technologically ambitious and economically expensive undertaking in human history. It would mark the end of human vulnerability to the environment, and the beginning of human ability to direct environmental change.

But public engagement, romantic notions of exploration, and future visions of life on Mars only give us a partial view of the context of twenty-first-century Mars exploration. In the space race, political support for planetary missions was driven by Cold War competition, as well as military and national security concerns that were translated into the vague notion of "leadership" in space. These factors remain important in the post–Cold War world, as space continues to be a place for spying and potential conflict with other nations. But the intervening years have also seen space connected to nearly every aspect of twenty-first-century life—to the point that many now fear how communications and global supply lines, not to mention military operations, would be disrupted by naturally occurring solar storms or attacks by weaponized enemy satellites. Space is very present in our lives, although mostly invisible. Mars exploration is just one way in which space, and what we are capable of doing there, becomes visible.

While its Mars program constitutes only a small part of NASA's activities, Mars exploration is one of the agency's most visible assets to the public and political world.[7] As Carl Pilcher, the former manager of NASA's solar system exploration program, once noted, "Mars is 'magic' in that it holds a special place in the minds of Congress, OMB [the executive Office of Management and Budget], and the public."[8] This explains why, despite the not infrequent cancellation of long-term Mars exploration plans, missions to Mars have nonetheless continued. There has been a kind of stability amid chaos for Mars scientists—the death of one plan is often overlapped by several competing proposals for a new plan, one of which will ultimately find traction. This is why from the outside it can appear to most observers that post–Cold War Mars exploration has been a coherent process that has progressively advanced our scientific understanding of Mars. It is also why in the present moment NASA can construct a narrative of how human exploration of Mars is the logical next step not only for Mars exploration but for its primary human spaceflight program—the International Space Station—in which years of lingering in low Earth orbit can be framed as continuous experimentation with living in the space environment leading to sustained presence on the Moon and Mars. The idea of embarking on such a grand mission has been the refrain of multiple presidential administrations from the 1980s to the present.[9]

For their part, scientists involved in Mars exploration know that funding for their work could dry up at any moment. Senior members of this community remember the rocky period between Viking and the first missions in the 1990s that restarted Mars exploration. The future of Mars exploration has often seemed to be in flux. Failures at Mars have had serious political consequences. Still, Mars scientists have come to expect—based on the history of their discipline, short as it may be—that Mars exploration is likely to continue. They recognize that it has to be tied in some way to politics, and when called upon, they advocate for the continuance of Mars exploration in congressional hearings. They embrace public outreach and educational initiatives that put Mars at the front of public consciousness and tie it to national leadership in science and technology. And they typically don't shy away from attaching their work to romantic notions of human exploration and expansion to other worlds.

Beginning in the 1990s and ending with the Mars 2020 rover mission, Perseverance, the United States has flown ten successful missions to Mars, each with a long and convoluted history that began long before it reached the launchpad. Each mission has been extraordinary, and scientists covet opportunities just to participate in, let alone lead, these endeavors. And yet Mars exploration has become such a "normal" activity within the landscape of twenty-first-century technoscience that robotic Mars missions have been the research site for no fewer than four anthropological studies of how the social process of science generates new knowledge about Mars. These studies, more than the mission websites and social media channels, present the humans who undertake Mars exploration. They make visible the many scientists and engineers (hundreds or thousands, depending on how you count) and the organized teamwork that makes the rovers work. They make it clear that the peculiarities of large teams working with robot colleagues has reshaped exploration and instituted robotic ways of working at sites like JPL.

William J. Clancey's *Working on Mars* demonstrated how, though rovers might be described as "robotic geologists," the process of operating them is far from the traditional experience of the field geologist. Because the rover moves slowly and time is at a premium, every rover action must be framed as testing a hypothesis, and team consensus must be established. When working with rovers, scientists come to identify simultaneously with the rover and with their fellow team members—a communal experience so immersive that when they say "we" they are referring to the rover, and when they say "the rover" they are referring to the team.[10] Janet Vertesi's book *Seeing Like a Rover* examines the social processes through which rover images are produced, manipulated, and interpreted to produce meaningful information and arguments about the geology of Mars.[11] Lisa Messeri's book *Placing Outer Space* emphasizes the ways in which scientific practices related to Mars and exoplanet science construct these planets as worlds—as literal places that humans can see and explore, even if only virtually.[12] Finally, Zara Mirmalek's *Making Time on Mars* examines the temporality of Mars-JPL as a dual-world work site, and proposes that we see the mission as a hybrid of earlier Arctic expeditions, with treacherous terrain and inclement weather, and "a stopwatch-driven factory production site, with a rigid timetable for task completion, decision-making, and social interaction."[13] As all of these works emphasize, Mars missions require a community of highly skilled and cooperative experts within a specialized work environment who have learned over time to manage the technological development and operation of the rovers and give them "life."

### A RETURN TO ORBIT AND THE FIRST WHEELS ON THE GROUND

The extended period of inactivity after Viking has generally been considered a pause in Mars exploration. In the words of JPL's then director Lew Allen, the 1980s were "a dry hole" for planetary exploration.[14] No US Mars missions made it to the launchpad until 1992. Finally on September 25, 1992, after a series of frustrating delays, NASA launched Mars Observer on a Titan III rocket. With more instruments than any previous orbiting mission—instruments tuned to data sets related to surface mineralogy, topography, magnetism, volatiles, dust, and the atmosphere—the spacecraft was essentially a flying laboratory, at a cost of $484 million. Meant to usher in a new era of less costly Mars exploration, this one orbiter in fact cost almost as much as the two Viking orbiters combined.[15] The launch and flight went perfectly until the spacecraft neared Mars. Three days before its intended entry into orbit, the team instructed the spacecraft to pressurize its fuel system. They never heard from it again. A review determined that it had most likely been crippled by an explosion upon pressurization.

Fortunately, Observer's failure was not the end of Mars exploration. NASA decided to refly the instruments selected for Observer, recovering its science mission on two follow-on missions that became known as Mars Global Surveyor and Mars Climate Orbiter, to be launched in 1996 and 1998. This new Surveyor model was in keeping with then NASA administrator Dan Goldin's new mandate for "faster, better, cheaper" missions that were more frequent but smaller and less expensive. Selected to fly on the mission were an infrared spectrometer, a high-resolution camera, a laser altimeter, a magnetometer, and an instrument used to measure variations in Mars's gravitational field. The overall goal of the Surveyor mission was to produce a global portrait of Mars over a full Martian year (the equivalent of two Earth years).

NASA launched Mars Global Surveyor in late 1996, as planned. Problems arose when the spacecraft arrived at Mars. To save fuel and weight, Surveyor had been programmed to slow itself not with rocket engines, but by dipping into the planet's upper atmosphere in "aerobraking" maneuvers meant ultimately to position the spacecraft in an optimal orbit. A structural fault in one of the solar wings caused it to bend during these maneuvers, and a four-month period of aerobraking was extended. Although the orbiter arrived at Mars in September 1997, its primary mapping mission did not get underway until April 1999. The science teams were able to make some observations from a higher orbit during the extended aerobraking, but this was mainly atmospheric data—mapping and mineralogy would have to wait.

When spectral analysis did become possible, it indicated that the planet was more complex than the Vikings had indicated. It was covered not in uniform volcanic material, but in two different types of volcanic rock: basalt and andesite. By creating a mineralogical map of the locations of these two types of rock and overlaying it onto photographic data and data from the laser altimeter, the science teams were able to create regional distribution maps of the two types of rock. The regional mineralogical variations matched up with the morphological differences that defined the ancient cratered highlands and the younger volcanic plains.[16]

Meanwhile, the new view of Mars provided by the camera system was raising more questions about the history of Mars than it answered. These images, on average about fifty times more detailed than the Viking images, showed gullies and sinuous channels in the sides of crater walls—features resembling gully washes found in the deserts of the American West. Images of the vast Valles Marineris canyon system revealed what looked like layers of alternating light and dark material, indicating that water on Mars had a more complicated history than Mariner 9 or Viking images had led geologists to believe. Some Mars scientists speculated on the basis of these and other features that Mars had once been covered in lakes and even seas at some time in its deep past. As the camera system's principal investigator, Mike Malin, told *National Geographic*, ". . . It all points to a Mars that was substantially more dynamic in terms of its environment, weathering, erosion, and transport than anything we see on the surface today."[17] Water had been active on Mars, but it was proving difficult for scientists to explain exactly how. The infrared spectrometer team looked for minerals that spoke to the history of water on Mars. The only such mineral they could identify from orbit was hematite, an iron oxide that forms in water—and this only in two locations on Mars.[18]

Another major triumph occurred at Mars during this period. Just weeks after Mars Global Surveyor left the launchpad, JPL sent its first rover, a small technology demonstration called Sojourner (named after the abolitionist Sojourner Truth) on its way to Mars. And on July 4, 1997, the "microrover" landed on Mars, sheltered inside the Pathfinder lander, itself a scaled-back version of a previously proposed Mars lander mission. Pathfinder and Sojourner were part of NASA's new Discovery program, meant to perform science on a budget with low-cost spacecraft focused on limited goals.[19] Pathfinder, in fact, did a relatively modest amount of science; and with the cost of the rover included, it didn't quite stay under the cost limit of the Discovery program. It did demonstrate, however, that a lander could reach the surface of Mars safely using an airbag hard-landing system that allowed it to survive crashing and bouncing around the surface. It also proved that a semi-autonomous rover could be operated from Earth. The rover was not primarily science-equipped; it only carried cameras and an alpha particle X-ray spectrometer (APXS) mounted on the rover's body. But for the first time, scientists were able to drive the rover up to rocks around the Pathfinder landing site and analyze the chemical composition of their surfaces with the APXS. The demonstration provided unequivocal evidence of what geologists had been proposing since the 1970s: that geology could be done on the surface of Mars with a properly instrumented mobile platform, a "robot geologist."

The Pathfinder-Sojourner development team knew that it had to use the mission to impress the public. It wasn't enough to just land on Mars again, especially if the smaller mission would do less than Viking had done twenty years earlier. They designed the rover to do just enough to show its potential, and to demonstrate that NASA and JPL were still innovating new technologies.[20] They were helped in this regard by the fact that computer chips and other hardware had become smaller, more powerful, and more affordable. Pathfinder could use a commercial modem to communicate with Sojourner while it roved on Mars.[21] But it was modems connected to home computers around the United States that made the Pathfinder-Sojourner demonstration a success.

In anticipation of the mission, JPL had set up a website for Pathfinder in 1996 with the plan to post new images on the site immediately after the lander's daily data return. After the landing, the website set new records for the relatively new World Wide Web, with close to forty million hits a day during the mission's first week.[22] The mission was defined as being only 10 percent science, but the scientists wanted to be able to test their APXS instrument on as many rocks as possible (plate 6). For their landing site they selected a location described as a "grab bag"—an outwash plain in Ares Vallis where water would have carried rocks from a variety of original locations. Web-savvy fans tuned in daily to see which rocks the little rover would visit next. It was a level of engagement nearly on par with that of the Apollo Moon landings. And it came as a surprise, given the small scale of the Pathfinder mission and the fact that the only humans involved were a bunch of rock-happy geologists and engineers. But the public was ready to geek out. NASA took note.

## BACK TO SQUARE ONE: FOLLOW THE WATER

The first post-Observer recovery mission was a triumph. The second Surveyor mission, Mars Climate Orbiter, had no such luck. NASA lost the spacecraft upon arrival at Mars on September 23, 1999. An unfortunate substitution of English units for metric in a crucial piece of software caused the spacecraft to either crash into the Martian atmosphere and burn up, or blow past Mars and into orbit around the Sun. Part of the "faster, better, cheaper" model of spacecraft development included lowering costs by performing fewer tests, and in this case it led to disaster. This was a devastating blow for those scientists who had now seen their instruments disappear on two ill-fated missions. Only a few months later, the Mars Polar Lander crashed into Ultimi Scopuli, a region near Mars's south pole. Adding these losses to that of the Mars Observer, JPL had now lost more Mars missions in the 1990s than it had delivered. These failures increased pressure on the lab to succeed at all costs. JPL now struggled to find an equilibrium between the streamlined work of "faster, better, cheaper" that had worked well for MGS and Pathfinder, and the traditionally reliable but more costly and time-consuming systems engineering approach developed during the Mariner era.

The Mars program itself still had the success of Mars Global Surveyor to show for its efforts; after all, it had discovered gullies and minerals related to water. JPL would have a chance to redeem itself again. In 1997 the lab had started developing a third Surveyor-type mission to be launched in 2001, which came to be known as 2001 Mars Odyssey, named after Stanley Kubrick and Arthur C. Clarke's 1968 film. This mission would continue NASA's investigation of the history of water on the planet, as NASA's Mars program had by now adopted the motto "Follow the water." A gamma ray spectrometer would search for subsurface ice, and a new thermal infrared imaging system would allow for finer-scale mineralogical mapping. The orbiter would also carry a communications relay to support a new rover mission—this time with a suite of scientific instruments. A future of human exploration once again became a long-term goal for NASA: an energetic particle spectrometer on the orbiter would measure the radiation environment on Mars to see what danger it would pose to human explorers, and an experiment on the rover would attempt to demonstrate that a human mission could make its own fuel from the Martian atmosphere, lowering the amount of weight the mission would have to carry with it from Earth.

The exciting water-related discoveries on Mars, combined with new Mars discoveries on Earth, put the question of life back on the table. In 1996, research on a meteorite found in the Allen Hills of Antarctica more than ten years earlier seemed to suggest that microbial life had in fact existed in Mars's deep past. The meteorite's chemistry indicated that it was part of a rock, formed billions of years in the past, that after being ejected from Mars several million years ago had ultimately come to rest on Earth thirteen thousand years ago. Scientists studying the rock made a shocking discovery: the meteorite seemed to hold fossilized evidence of Martian bacteria. Intense scrutiny of the rock over the next two years by the scientific community picked apart the claims of life, but the very public nature of the announcement and subsequent discussion piqued popular enthusiasm for the search for life on Mars, and it got the scientific community thinking again about how to look for life—including evidence of past life—on the planet's surface.[23]

A discovery from twenty years earlier seemed as though it might hold the key to rethinking life on Mars: oceanographers using submarines to study the newfound phenomenon of hydrothermal vents in the Pacific Ocean were surprised to find entire ecosystems thriving in complete darkness, sustained by the chemistry of the vent itself. These newly discovered organisms, called chemolithotrophs, joined another group of microbes that scientists were calling "extremophiles" (life that thrives in extreme environments) as further evidence that there was hardly a place on Earth where one couldn't find microbial life.[24] The one thing that seemed to be required in all instances was water.

The next Mars rover mission would be delayed until 2003, and the fuel experiment would not fly until 2020. Odyssey launched as planned in the spring of 2001, reaching Mars orbit on October 24 of that year. By early 2002, the gamma ray spectrometer team reported that they had finally answered one question about the water on Mars: that of where it was hiding. The water was practically everywhere, bound up in ice beneath the surface and blanketed around the planet in red, dusty permafrost. The infrared imaging system directly targeted areas of interest where liquid water seemed to have once been present. Mineralogical maps of Mars based on infrared data grew more sophisticated, providing information about the distribution of a large variety of igneous rocks and minerals. The new data sets represented the first time in the history of Mars exploration that overlapping spectral and visual data had been collected at resolutions useful for studying local geologic processes; they could identify the composition of older parts of the crust, uncovered by local erosion and impact events. Geologists were beginning to see a Mars that was dominated by volcanic basalt but had interesting regional differences.

In 2000, JPL had started developing a new robotic "Mars geologist" rover to follow on the success of Sojourner. Ideas from various projects came together into what became known as the Mars Exploration Rover (MER), a compact robot about the size of a golf cart that could be folded inside a modified Pathfinder landing system (plate 7). Like Pathfinder, the MER would crash onto the surface of Mars, bounce on airbags until it came to rest, and unfold safely at its landing site. JPL would build two rovers instead of one, a decision that potentially doubled the amount of science the mission could accomplish while lessening the potential political damage of losing one of the rovers during the risky landing procedure. Within JPL, the MER mission combined the younger, streamlined, risk-taking team that had made Pathfinder work under Goldin's "faster, better, cheaper" mandate with JPL's more traditional systems-engineering-oriented staff. As the project was larger and more ambitious than Pathfinder, it grew to absorb a large number of JPL staff.[25]

The two MER rovers, named Spirit and Opportunity, were launched on June 10 and July 7, 2003. They arrived at Mars three weeks apart in January 2004. Both survived the terrifying landing sequence. Opportunity was sent to Meridiani Planum, a cratered plain where infrared data indicated a large exposed patch of hematite (figure 18). Once on the ground, it found small round spheres that the team nicknamed "blueberries" due to their shape and the blue hue they took on in some false-color images used for analysis (figure 19). Analysis of the blueberries confirmed that they were made of hematite.[26] The blueberries—along with other minerals formed in water, sandstone layers cemented by sulfates, and shapes etched in the rock—spoke to a local history of repeated flooding, wet volcanic

**FIGURE 18** A Mars Exploration Rover examines the Martian soil with an instrument-bearing robotic arm. NASA/JPL.

FIGURE 19 "Blueberries" at the Opportunity landing site provide evidence of a wet past. NASA/JPL.

muds, and evaporation. This was the first ground truth found on the surface of Mars that confirmed a hypothesis derived from orbital data.

Spirit's landing site was chosen based on its visual appearance. It was a crater named Gusev, which Steve Squyres, the science lead of the MER mission, believed might once have held a lake. Once Spirit was on the ground, this hypothesis seemed at first to be disproven. The minerals in the rocks and dirt that filled the plains of Gusev showed only minimal interaction with water, and they were all basaltic lavas. This provided a valuable lesson about the deceptive nature of resemblances between Martian and terrestrial surface features: Things aren't always what they appear to be. Squyres and his team decided to move on, commanding the rover to drive to a large outcrop of rocks named the Columbia Hills (after the lost crew of the space shuttle Columbia) about two and a half kilometers from the landing site. The scientists were sure that water must have interacted with this region, and this outcrop seemed like a better place to find it than on the lava-covered floor of the crater. Here they found a variety of rocks that might have been altered by water. After Spirit died on March 22, 2010, analysis of data from the rover's onboard spectrometer suggested that two outcrops in particular, nicknamed Comanche and Comanche Spur, had once been soaked in water filled with carbonates—possibly from a hydrothermal source.[27]

### EXPLORATION CONTINUES

As explained in the introduction to this chapter, the Spirit and Opportunity rovers survived much longer than expected, and in the process were able to cover much more ground and do a lot more science than originally anticipated. Spirit traveled five miles over its seven years on the planet, while Opportunity covered just over twenty-eight miles over fourteen years. Graduate students who began their careers working with these rovers were mid-career professors, training students of their own, by the time the mission ended. A third generation of Mars explorers had now come of age. At JPL, operating the twin rovers required more personnel and person-hours over a longer period of time than any previous mission. While there was certainly a learning curve in figuring out how to keep the rovers going without burning out staff and scientists, this did little to diminish enthusiasm for future lander and rover missions.

Mars exploration continued. As this book goes to press, Mars Odyssey is still operating and has moved into an orbit that allows its infrared imaging system to continue its mineralogical mapping of Mars. The Mars Reconnaissance Orbiter—with the highest-resolution camera system ever sent to Mars, the telescopic high-resolution imaging science experiment (HiRISE)—arrived at the planet in 2006 and has been operating with the express purpose of locating and characterizing potential landing sites and scouting out regions for rovers to explore. In 2008 the

Phoenix lander arrived at Mars. When the thrusters of its soft-landing system fired, they cleared a layer of Martian dust at the landing site and revealed what turned out to be water ice—the first such sighting by a spacecraft (figure 20). Analysis by the lander's chemistry experiment, essentially the first miniaturized chemistry lab sent to Mars, confirmed that there was indeed water ice at the landing site. NASA had followed the water, and had found it.[28]

Then came the Mars Science Laboratory mission, with the Curiosity rover—the largest and most robustly instrumented rover yet (plate 8). Unlike the MER rovers, which carried the tools of a field geologist, Curiosity was a mobile science platform the size of a small car. It carried instruments and experiments related to geology, geochemistry, biology, atmospheric and hydrological science, and radiation. It also outdid the Sojourner and MER rovers in the movement and functionality of its robotic arm, which not only can use a suite of attached instruments to analyze rocks, but also can deliver samples to the experiments carried within the rover's chassis. Launched on November 26, 2011, Curiosity landed in Gale crater on August 6, 2012. The landing site, a ninety-six-mile-wide impact crater, was selected on the basis of infrared data that indicated the crater might once have held a large lake. At the site, science teams using Curiosity's instruments confirmed that the floor of the crater contained clays and sulfates. The rover spent several years ascending a three-mile-tall mound of sedimentary debris nicknamed "Mount Sharp" (officially named Aeolis Mons) and studying its mineralogy one layer at a time. The science teams speculated that the base of the mound was the remnant of sediment laid down over perhaps as many as two billion years by a lake that once filled the crater. Some have proposed that the lake was a temporary but recurring body of water created by a series of flash floods.[29] The minerals and their presence in Mount Sharp speak to a time when Mars's climate was saturated with water. The science teams are still trying to get these layers to tell them when and why Mars became the dry world we know today.

Other spacecraft have followed. In 2014, NASA's Mars Atmosphere and Volatile Evolution (MAVEN) orbiter arrived at Mars and began its

**FIGURE 20** Ice on the Martian surface, found by the Phoenix lander. NASA/JPL.

mission of studying the planet's upper atmosphere and ionosphere in an attempt to understand their relationship with the solar wind. Another NASA lander, InSight, arrived in late 2018, carrying instruments to detect Marsquakes and subsurface temperatures. Most recently, in February 2021 NASA's Perseverance rover landed in Jezero crater, another site on Mars suspected to have held an ancient lake (plate 9). The feature that most attracted scientists when selecting this landing site was what appears to be a river delta where sediments from water flowing into a lake within the crater were deposited. Perseverance, similar to Curiosity in many ways, is specifically tuned to look for biosignatures—signs that life might once have enjoyed Mars's warmer, wetter past. For the first time, it carries a drill that can collect rock core samples that NASA plans to retrieve and send back to Earth in a series of future robotic missions. It also carried a successful technology demonstration: a small autonomous rotorcraft or Mars helicopter named Ingenuity (plate 10), which captured the imagination of a world still in the throes of a pandemic. If the history that developed in the wake of the Pathfinder mission is any indication, we may see increasingly capable Mars helicopters sent to Mars in the future.

And of course, although NASA has had the most success in sending spacecraft to Mars, the United States is not alone in its interest in the red planet. After the end of the Cold War, Russia tried and failed again to reach Mars with its Mars 96 mission. Japan lost contact with its Nozomi spacecraft in 1998. The European Space Agency succeeded in putting its Mars Express spacecraft in orbit around Mars in 2004, and it is still operational. But the ESA lost the Beagle 2 lander deployed by the orbiter. A joint Russian-Chinese attempt to reach Mars in 2011 failed to get past Earth orbit. In 2013, India's Mangalyaan orbiter, also known as the Mars Orbiter Mission (MOM), succeeded in reaching Mars, where it remains operational. The year 2016 saw the successful arrival of the ESA-Roscosmos ExoMars Trace Gas Orbiter, along with the loss of the accompanying Schiaparelli lander. Finally, in 2021 the Emirates Mars Mission delivered the Hope orbiter to Mars, and China successfully delivered the Tianwen 1 orbiter and the Zhurong rover. Exploration of Mars is becoming an increasingly global enterprise.

**THE AGE OF THE MARS EXPLORER: NERDS IN LOVE**

If the above story of Mars exploration seems more technical than in previous chapters, this reflects the reality of Mars science today. After Viking, the era of attempting to deduce the characteristics of Mars from physical principles and speculation about the original conditions of the solar nebula gave way to an era of esoteric geological exploration. Detailed visual and spectroscopic data of surface features allowed geologists to begin working to determine the processes that had formed them. This was a new mode for Mars science, in which "the landscape itself . . . serves in the role of the 'experiment'" from which a history of the planet must be reconstructed.[30] The present conditions of Mars became the key to the past. Questions about Mars were to be posed and hypotheses articulated in the language of surface morphology, crater counts, stratigraphy, and geochemical composition. Finding answers to these questions involved engineering new instruments, sifting through piles of overlapping data sets, and navigating the vagaries of a political and bureaucratic government agency.

In previous chapters we focused on how Mars was manifested in the human imagination, and how the imagined Mars overlapped with what people believed about the universe in which they lived. In the twenty-first century, the histories of Earth and Mars have come into even closer alignment, and robotic landers and rovers allow humans to replay the history of terrestrial exploration on another world. These spacecraft are tools, of course—not explorers in their own right. The explorers are on Earth, in large teams of scientists and engineers from universities and research centers around the world. So before we look to works of fiction to see how Mars exploration is presented, we should look to the accounts of those who are today traversing the red wastelands of Mars with robotic assistance.

Who are these Mars explorers? As Clayton R. Koppes notes in his history of JPL, the lab was primarily "a white male preserve" through the 1970s.[31] NASA in general had the reputation of being, in Goldin's words, "pale, male, and stale." In the 1980s, however, the lab's personnel began to turn over, as senior engineers retired and were replaced by a new generation. The new cohort remained largely white and male, but it also represented new talent from Asia and the Middle East, including the Lebanese-born Charles Elachi, who later went on to head the lab. Women began to make up a larger segment of the JPL workforce; and from 1984 to 1994, women climbed from 9 to 15 percent of the science and engineering staff.[32] Within the story of Mars exploration, Donna Shirley, who joined JPL in 1966 as one of its few women engineers (and the only one with an engineering degree), served as development lead of the Pathfinder mission and then went on to become the manager of the lab's Mars Exploration Program. In 1998 Shirley published *Managing Martians*, a memoir of her time at JPL, which describes her path from engineering student to leader of a team of thirty engineers designing Sojourner and making the rover a reality.[33]

Shirley's book may have been among the first of what we can call Mars memoirs, which have become popular in the years since Pathfinder. Publishers took note of the public interest in Mars landers and rovers, and participants in these missions found literary agents and editors who were happy to entertain book proposals about Mars exploration. The charismatic scientists who managed to catch the public's eye in the 1990s put forward an image of Mars exploration that was a far cry from the "cool militarism of the old Cold War space program" and instead presented the new model of "nerds in love."[34] The firsthand accounts of Mars exploration published in the last quarter century of robotic exploration certainly do convey an unbridled enthusiasm for Mars as a site for asking and answering questions, or as a field on which to test new technologies. While written in the first person, these books often emphasize the teamwork and human relationships that make robotic missions work.

Many of these memoirs invite readers behind the curtain to show the human side of exploration. Andrew Mishkin's *Sojourner: An Insider's View of the Mars Pathfinder Mission* (2003), promises a behind-the-scenes look at JPL, the lab that does "what's never been done" and goes "into the solar system where humans cannot yet go."[35] Steve Squyres's *Roving Mars: Spirit, Opportunity, and the Exploration of the Red Planet* (2006), written just two years into a mission that lasted more than a decade, spends twelve chapters going from idea to proposal to the convoluted process through which the MER program eventually was born, and finally to the spacecrafts' development and launch. The remaining five chapters finally get us roving on the red planet. No doubt this accurately represents Squyres's experience as the main force pushing for an instrumented science rover. But Squyres also tries to tap into the mythos of the explorer. Mars became his target, he explains on page 1 of his prologue, because he wanted to explore, but "there were no blank spots on the maps anymore."[36] Like many in his generation of Mars explorers, Squyres was pushed toward Mars by circumstance. He happened to be an undergraduate at Cornell University during the Viking mission, and he serendipitously discovered that a geology course would be taught by one of the project's participating scientists. Over the course of his undergraduate and graduate education, he developed a close working relationship with Carl Sagan, who helped to shape him into a planetary scientist. Squyres's book tells us how MER got to Mars, and gives us a window into the first years of exploration; but even more, it gives us a portrait of the career of one of the most successful members of the second generation of robotic Mars explorers.

The first pages of Rob Manning's *Mars Rover Curiosity: An Inside Account from Curiosity's Chief Engineer* (2014), play up the drama in the "seven minutes of terror" of the rover's entry, descent, and landing on Mars. But ultimately, Manning's is a story of solving design challenges. He conveys suspense in the statement "If everything works as designed"—which, often enough, it hasn't. Manning describes the packed control room, full of "team members anxiously huddled at their displays," trying to wish away all the things that could go wrong. Ultimately, this is a story of people sitting in front of computers. The drama is in the design, and in the science and technology used to explore another world.[37]

At their best, Mars memoirs convey a level of enthusiasm about exploration and the production of knowledge that is difficult to find in more academic histories or ethnographies of Mars exploration. They're also much more personal than those academic accounts. The astrobiologist Sarah Stewart Johnson's *The Sirens of Mars: Searching for Life on Another World* (2020) delivers a well-curated chronology of previous eras of Mars exploration, and recounts the stories of those whose passions and research have informed her own thinking about the red planet, as she applies the still young science of extremophiles to the question of life on Mars. The people she describes in her book are kindred spirits: some are from past eras, peering at the fuzzy red disk of Mars through observatory telescopes or from hot-air balloon gondolas, and others are contemporaries working alongside her in university laboratories or at far-flung field sites. In their stories she finds "the great strides forward, the longing for answers."[38] Johnson's account is one of a participant in history, of being among the first cohort of women to explore Mars. In *The Sirens of Mars*, the excitement and the occasional crushing disappointment of Mars science become personal.

### TWENTY-FIRST-CENTURY MARS IMAGININGS

We continue to imagine a future in which humans will explore Mars, bringing their earthly bodies with them; and futures in which humans will even live on Mars and beyond (plate 11). Because real Mars images are always and everywhere available in this digital age, these imaginings tend to be based in the Mars landscape we've come to know (plate 12). Audiences are not likely to appreciate Mars science fiction that doesn't engage with the harshness of the Mars environment and the hardships human explorers or settlers would have to face. Mars may look like Earth, but it is one of the most hostile environments one can imagine; even the dust wants to kill you. The Mars explorers of the millennial imagination don't have it easy.

One of the most popular recent examples of an imagined Mars journey is Andy Weir's 2011 novel *The Martian*, along with Ridley Scott's 2015 film adaptation. The novel, Weir's first, recounts the fictional exploits of an astronaut left for dead on the surface of Mars with only a small and dwindling supply of the resources he needs to survive. The very first sentence of the novel sums up the difficulty of surviving such a mishap on Mars: "I'm pretty much fucked."[39] But Weir, like his protagonist Mark Watney, is a nerd. Primarily using Internet sources, he was able to think through the technology and resources that would likely be available to Watney, and find plausible (though unlikely) survival strategies. Not even able to communicate back to Earth, Watney must survive using only his own expertise in engineering and botany. He has the mission's habitat to shelter him, but his solutions to problems often require him to travel great distances over the Martian surface and engineer new, more mobile shelters on the fly. Watney also uses some of the robotic tools left on the surface of Mars during previous eras of exploration—including Pathfinder, which he repurposes into a means of communication with NASA. Watney's now famous line from his video journal, "I'm going to have to science the shit out of this," is very different from what we might find in a John Carter story, where chivalry and violence are the key to survival on Mars. Much as in the real stories of robotic exploration, Watney's survival depends on his ability to solve intractable problems with innovative solutions. It's "nerds in love," with higher stakes.

Harrowing survival stories like *The Martian* draw readers and audiences. They increase excitement about crewed missions to Mars and make the necessary technologies (futuristic suits, habitats, and roving vehicles) visible. But they also make it difficult to imagine what the purpose would be of sending humans to such a hostile world—other than resetting the goalposts of human exploration for exploration's sake. The tech is all very cool, and the scenery is breathtaking, but what makes it worthwhile? It's not at all clear from Watney's story what Earth's long-term goals are on Mars, nor how the political will and economic capital have been mustered for such a venture. But we are led to believe that human Mars exploration became an enterprise in which multiple nations were willing to invest.

Other future visions show humans not just exploring but living and thriving on Mars and beyond. As in *The Martian*, it's often not clear in these stories how or why humanity overcame the technological and economic obstacles that now prevent such endeavors. Typically, the stories involve new technologies that are not incremental improvements on what we use today, but de novo technologies that have changed the nature of spaceflight entirely. One popular version of this vision is *The Expanse*, a series of novels written by Daniel Abraham and Ty Franck under the pen name James S. A. Corey. The series has also been adapted into a streaming video series for Amazon Prime. It depicts what we might consider a mature spacefaring human species that has populated not only Mars but parts of the asteroid belt and moons of the outer planets. On one level, the depictions of humans in spacesuits and of spaceships obeying the laws of physics are very realistic. There is also a somewhat compelling reason for humans to want to live on Mars: Earth has become dangerously overpopulated.[40]

But beneath this, the technologies that drive these ships and allow large-scale habitability on the Moon, Mars, and Ceres require readers and viewers to accept the conceit that yet unknown science (fusion drives, "Epstein" drives, stroke-preventing "crash couches," etc.) will someday soon make spaceflight safe and affordable and put Mars within reach. In doing so, of course, stories like *The Expanse* draw upon a long tradition in science fiction of imagining technological placeholders that permit speculation about how space will change us. The "Dusters" (Martians) and "Belters" of *The Expanse* no longer feel any allegiance to Earth; they are human, but generations of living on other worlds have made them culturally and even physiologically distinct. Grand terraforming initiatives have stalled, as new generations of Martians, accustomed to living in large domed habitats, find it unnecessary to make Mars more like Earth. Like some of their 1960s counterparts, Abraham and Franck imagine Mars as a hypercapitalist, militaristic world, where a new society has been forged through a process defined by an unequal colonial relationship with Earth, an ensuing civil war for independence, and the hardships of life on an inhospitable planet. The Martians we imagine today are tough, to say the least.

Just as the population of real Mars explorers has become more diverse over the past few decades, so too have the characters who inhabit imagined Mars futures. *The Expanse* boasts one of the most diverse casts in the history of American science fiction television, and the novels on which the show is based present a multiracial and ethnically diverse population occupying all parts of the solar system. One might be inclined to call the books postracial, in that they depict a future in which race has ceased to be a meaningful category, and has become instead only a matter of appearance. However, the prejudices between Earthers, Dusters, and Belters is clearly an analogy for the racism and classism that exist in our world.

Other contemporary works have examined how race, ethnicity, and class as they exist today will continue to shape who benefits from spaceflight, including trips to Mars. Silvia Moreno-Garcia's 2017 novella *Prime Meridian* recounts the story of Amelia, a young woman in Mexico City who dreams of traveling to Mars. In this future, there are two ways to get to Mars. If you can find a company to sponsor you, you can get a class C visa that makes you a perpetually indentured worker. But Amelia wants to use hydroponics to grow genetically modified plants on Mars, which would require her to invest in Mars by paying for a coveted class B visa.

Moreno-Garcia, who has also written quite a few gothic horror novels, is no stranger to vampires. There are no literal vampires in *Prime Meridian*, but there are people who pay Amelia for her friendship, and even for her blood, which they use in rejuvenation treatments. This is, in a way, a microcosm of the larger vampire at work in Amelia's story: resources are being poured into Mars colonization even as the Earth economy is broken, failing those, like Amelia in Mexico City, who live in poorer regions and whose lives have become precarious. The choice of who gets to go to Mars is defined by the same forces that maintain wealth inequality on Earth. This is not the story of a girl whose dream of Mars comes true; it is the story of a girl whose dreams of Mars sustain her in her times of hopelessness. When Amelia sells her blood, she lays back in a recliner, as if in a spaceship bound for Mars. She stares up at the ceiling and tries to imagine Mars, which doesn't always manifest.[41] Moreno-Garcia's work reminds us that solving the technological problems of getting to Mars and surviving there will not create a utopia, especially if we have not protected the lives of everyone on Earth.

### FURTHER AHEAD ON MARS

Toward the end of the first quarter of the twenty-first century, at least two futures for Mars seem to lie ahead. One path leads to a Mars that remains distant from human activity. Scientists and engineers will likely find new ways of exploring this Mars and drawing from it the secrets of its past. They will send follow-on missions to Perseverance that will launch and bring home the rock samples the rover collected. Perhaps they will find some form of Martian life, or evidence that it once existed. Perhaps they will uncover planetary knowledge or new technologies that will help abate or alleviate the impacts of climate change. It's even possible that this extended period of interest in Mars science will wane, replaced by missions to the icy ocean moons of the outer planets. All of this can happen without one human ever setting foot on Mars.

The other path leads to humans on Mars. We don't yet know exactly what this future will look like—whether it will be small-scale scientific expeditions, an orbiting Mars station, or full-scale efforts to build new communities or even cities on Mars. Perhaps all of these things can happen, and perhaps one will lead to the others. At present, however, there is no solid plan for such endeavors. It's possible that the success or failure of the currently planned Artemis missions to the Moon will at least partially determine the fate of a human Mars. Will NASA and its partners succeed in building a sustainable presence on the Moon? Or will we find ourselves living through another Apollo: another impressive technological spectacle that is abandoned when the goal is reached and the price has grown too high? There is no agreed-upon answer to the question of what ends these efforts at the Moon, let alone Mars, will serve. Who will benefit from humans on Mars? Will these journeys serve a larger purpose, such as developing new technologies that help us live better on Earth? Or will they be used to extract new wealth from a new environment? Will they lead to a utopia, a dystopia, or something in between?

CONCLUSION

# *THE HUMAN FUTURE OF MARS?*

There is a strong temptation, reaching back to the very dawn of the space age, to set Mars as a goal for human spaceflight. Among today's space enthusiasts, there is not only a nostalgia for the science fiction futurism that made humans living a multiplanet existence on Earth and Mars seem inevitable, but a sense that NASA and other government space agencies have failed to live up to the promise that human missions to Mars would follow the success of the Apollo lunar missions. This sentiment is shared by at least one Apollo astronaut, Buzz Aldrin, the second human to set foot on the Moon. Aldrin expressed his disappointment with the post-Apollo period of robotic exploration and Earth-orbiting space stations, insisting, "We've been stuck in low Earth orbit for too long and I believe that we need to break this malaise." Aldrin's own plans would have humans on Mars by 2039, the seventieth anniversary of his Apollo 11 Moon landing. He argues that this time, we should do it in such a way that we can sustain a long-term human presence on Mars, as opposed to the "flags and footprints" blip that Apollo became.[1]

This critique can also be found in popular culture. The television writer and producer Ronald D. Moore, whose experience with imagining the human future in space includes *Star Trek: The Next Generation*, *Star Trek: Deep Space Nine*, and the reboot of *Battlestar Galactica*, turned to counterfactual history in his 2019 Apple TV series *For All Mankind*. The timeline of *For All Mankind* diverges from our own in 1969, when a crewed Soviet spacecraft lands on the Moon between the Apollo 10 "dress rehearsal" flight and the Apollo 11 landing. This unexpected upset ensures that the Moon remains a Cold War battleground for the next two decades. Astronauts and cosmonauts build lunar bases and mining operations, and the two nations set their sights on Mars. As Moore explained in the show's synopsis, this timeline allowed him to enact "the space program that we were promised but never got."[2] He also decided to make the show optimistic, bucking what he saw as a dystopian trend in alternate history, and highlighting how "by expanding the space racing . . . the world became a better place, and the nation became a better place."[3] Ambitious space plans are good for us, his show implied. They're the fortifying vitamin American society has been lacking; the bigger the dose, the better.

Those of Aldrin's generation, as well as those like Moore, who watched the Apollo missions on television as children—and, for that matter, those of my own generation, who came of age amid the triumphs and tragedies of the space shuttle program and the International Space Station—have seen NASA administrators and presidential administrations push plans to send humans to Mars, only to have deadlines slip and ambitions fizzle out. The elusive goal of setting boots on the red planet always seems to be two or three decades in the future—close enough to justify funding the development of new technologies, but distant enough to permit our image of human Mars exploration to remain fuzzy. A multitude of plans have rushed in to fill this gap, each not quite agreeing with the others about justification of the expense, the purpose behind the risk to life (human or Martian), or even the technologies required.

For its part, NASA still seems to be playing a long game with Mars. Since 2015 the space agency has had the goal of returning to the Moon and then, having established a foothold there, embarking on a "journey to Mars." The current expression of this goal is NASA's Artemis program, which has the stated aim of landing the first woman and person of color on the Moon. In so doing, NASA plans to leverage commercial partners and international allies to construct a Lunar Gateway space station in orbit around the Moon, an Artemis base on the surface, and a Human Landing System for ferrying astronauts between the base and the station. It also plans to use one of the largest rockets ever built, the Space Launch System (SLS), and a state-of-the-art capsule called Orion for the trip to the Moon. But as I write this conclusion, the rocket is faring poorly in tests, and the projected cost of human launches to the Moon is still very high. Despite these setbacks, NASA administrator Bill Nelson insists that the program "will demonstrate NASA's commitment and capacity to extend humanity's presence on the Moon and beyond."[4]

Meanwhile, NASA's most recent Mars rover, Perseverance, is meant to pave the way for human exploration of Mars. Not only is it collecting rock samples to be sent back to Earth by follow-on robotic missions for laboratory analysis—work that will no doubt help in better characterizing the Martian environment, and in determining what will be required to keep humans alive there—but it also carries experiments directly related to human needs. One experiment, the Mars Oxygen In-Situ Resource Utilization Experiment (MOXIE), first proposed in the early 2000s, has shown that oxygen can be produced from the planet's carbon dioxide–rich atmosphere. Oxygen is, of course, required for humans to breathe, but it is also a component of rocket fuel that could be used for the return journey to Earth. Spacesuit materials can also be found on the Perseverance rover, and another experiment is recording how those materials hold up to the Martian elements over time. While these components don't add up to a fully realized map to Mars, one can imagine how each is an incremental step toward the planet.

NASA dangles Mars mission concepts in front of the public not only as a goal for human spaceflight, but as one of its ultimate purposes. If we are going to realize the dream of spaceflight, become a multiplanet species, and perhaps even reach the habitable Earthlike worlds that exoplanet research suggests exist elsewhere in our galaxy, then learning how to get to Mars and back is a necessary first step. If low Earth orbit has been a set of training wheels, then a mission to Mars might be our first ten speed. However, NASA, while it is one of the most trusted government agencies in the country, lacks the political and fiscal resources to get there. This has led to a confusing mixture of hype and restraint in Mars mission planning, which doesn't sit well with some of the agency's critics. It's safe to say that most Americans will probably react to Artemis, and to any Mars mission ideas that follow, similarly to how the previous generation reacted to Apollo: they will be inspired by the achievement, and may even buy into the interplanetary vision, but will be unsure of how their interests are served by the necessarily large expenditure of public funds. According to survey results, only about 20 percent of the American public believes that we should allocate more money on spaceflight. In the wake of Apollo, 40 percent felt the cost was too high.[5] Apollo has many historical lessons to teach us, and one of these is that sending humans to other worlds requires that a nation makes civilian spaceflight a top national priority—something very few Americans are willing to support, no matter how much they enjoy consuming popular depictions of Mars missions.

One way to get around this problem is to imagine that we can get to Mars without spending public money. The theoretical physicist and futurist Michio Kaku is excited about NASA's Artemis program and the "journey to Mars," as it means a return to investment in large launch systems like SLS and the SpaceX Starship rather than low-Earth-orbit infrastructure. He warns, however, that NASA's failure to set realistic goals for Mars exploration leaves the agency "rudderless," continuing along a path marked by "muddle, vacillation, and indecision."[6] If we do make it back to the Moon, or on to Mars, Kaku gives the credit to "the billionaires who are tired of the lumbering pace of NASA bureaucrats" and who want to see the pace of exploration quicken.[7]

Kaku is one in a very vocal subset of space enthusiasts who believe that the path to Mars will be charted not by the government but by commercial enterprise. For them, the problem is not that the government has spent too little on spaceflight, but that it has monopolized control over space and restricted access to it for too long. They cite the recent successes of private companies like SpaceX, Blue Origin, and Virgin Galactic—each backed by a billionaire founder—as evidence that the future of space is commercial. These space entrepreneurs have leveraged their vast wealth—none of which was earned in aerospace—against the promise that space can be profitable (beyond, but not excluding, the money made through contracts with space agencies, the military, and intelligence agencies). At least one of these billionaires has set his sights on Mars. Elon Musk, founder of SpaceX, plans to start sending missions to Mars in the next decade, using his company's super-heavy-lift Starship rockets, and he estimates that it will take a minimum of twenty-two years to build a self-sustaining settlement on Mars. He reportedly has promised himself to build a one-million-person Mars city by 2059, the year he turns seventy-nine.[8]

### WHY GO?

We can call Mars the most Earthlike planet in our solar system. And yet Mars remains distinctly uninhabitable—or at least very inhospitable. The surface is cold and constantly bombarded by radiation; the atmosphere is incredibly thin and made up mostly of carbon dioxide. The soil, if we can call it soil, is toxic to anything we would want to grow there, and the dust storms that occasionally cover the entire planet carry a very fine powder that would damage skin, eyes, and lungs. In short, for the first astronauts on a return trip to Mars, the red planet would be one of the most hazardous worksites ever attempted. For anyone who tried to live on the planet in the long term, Mars would present more challenges than the most remote and hostile places on Earth. And yet so much of the discourse surrounding Mars today treats the planet as a new frontier—a territory filled with nothing but wide-open spaces waiting to be transformed by the human hand.

But why? Assuming that humans can use technology to live on Mars as it is now, or to alter it into a more habitable planet, why would we want to do it? How would we benefit from putting humans on Mars?

If we know what our objectives are in going there, then we can more clearly calculate (even if in qualitatively fuzzy terms) what its value might be.

One common refrain in arguments for sending people to Mars is that humans have an innate urge to explore. For some spaceflight advocates, like Robert Zubrin, founder of the Mars Society, the urge to explore is a defining human trait. To not explore space, to not set our sights on other worlds, would be to become "less than human."[9] Zubrin's argument is based not only on the history of human migration and exploration, but on an understanding of vertebrate evolution. Humans are descended not only from other hominids but from the creatures who left the ocean, faced the hazards of the atmosphere for the first time, and learned to live in what was until then a threatening and even deadly environment. What is space but the next evolutionary horizon?

Assuming that there is, to paraphrase Zubrin, an extraterrestrial imperative written into our DNA, we can still ask the question of whether more good or harm would come from acting on this urge. Is it a good justification for sending humans to Mars? Will it yield good consequences? For Zubrin, the history of human exploration on Earth is the history of progress. It is through exploration and expansion that humans have learned to better themselves and better understand the world they occupy. But this narrow view of exploration and expansion privileges the perspectives of those who have benefited from these enterprises, at the expense of those who have been harmed. A more complete understanding recognizes that European western expansion and colonialism brought with it oppression, inequality, and genocide of Indigenous peoples. There may have been progress in science and technology, but not necessarily moral progress.[10]

Mars colonization advocates may ask whether this history is even relevant to their plans. After all, Mars is not Earth. There are no Indigenous Martians to encounter or displace. Yet the conceptual and linguistic framework with which they imagine and speak of putting humans on Mars invokes colonialism. The Mars colonists they imagine are usually there either to serve commercial ends, mining valuable natural resources, or to recreate versions of American frontier life. These plans also tend to include only one nation or group, and their goals require making claims on Mars and its land that will limit what others can do with it.[11] In other words, there may not be Martians to contend with now, but soon enough there will be. The cultural lens we apply to our future on Mars—colonial or, preferably, decolonial—will determine which politics and prejudices we bring with us. A Martian land rush would certainly recreate the problems of our past. If we are going to Mars, an increasing number of critics argue, we should abandon this way of thinking entirely. We should think in terms not of workers and extractive processes, but of more collective and equitable ways of living. We should imagine technologies and purposes that do not limit our ability to put humans and communities first.[12]

It seems to me that there are at least four things we should avoid in any endeavor to put humans on Mars. We should not extend the environmental problems we've created on Earth to Mars by bringing with us unsustainable industrial processes and ways of living. Nor should we use Mars to escape from those problems, especially if this escape is only available to a privileged few. We should not use Mars to replay colonization histories that exploit and exclude. Likewise, we should not create new inequities by developing Mars opportunities for some at the expense of others. I think this means that if we do go to Mars, we should do it in a way that maximizes the benefits to Earth. Exactly how this works is something we should discuss—and the "we" we bring to the table should be as inclusive as possible. This should go well beyond the limits as they are currently construed by the treaties and accords that govern space. As the journalist Adam Mann notes, "At the moment, the scales are tipped heavily in favor of people and nations that have been historically powerful."[13]

I also think we should begin this discussion by addressing two of the biggest problems that face our world today: climate change and wealth disparity. At minimum, any plan to send humans to Mars has to have a neutral carbon footprint and not amplify or reinforce economic inequality. At best, the technologies we develop to transform Mars or live sustainably on the red planet will be applicable to reducing emissions and/or removing carbon dioxide from Earth's atmosphere. On this end of the spectrum, any knowledge or wealth derived from Mars will help to lift all people, not just opportunistic investors or entrepreneurs. If Mars belongs to humans, then it belongs to all humans. Discussions of what to do with Mars should include as many voices as possible.

### WHO ARE WE NOW?

Another way to look at the question of what we should do on Mars is to rephrase it as a question about who we are to think about Mars today, and who we think we will become by making the journey to Mars. Can we unpack our assumptions about ourselves and our world, as we have unpacked our assumptions about the humans we've followed throughout this book? For a historian like myself, the present is not always easy to penetrate. But I can try.

Our ancient counterparts knew themselves to be participants in an animate world, with no clear divide between natural and human events. Medieval thinkers saw their universe as divinely ordered, and governed in a chain of connections originating from God. Early modern Europeans believed themselves to be at the center of an expanding world waiting to be encountered and discovered. In the eighteenth and nineteenth centuries, Europeans and Americans thought in terms of settling interior territories, structuring colonial relationships, and imposing order upon a changing world. The Cold War divided the world into spheres of influence, and seemed to necessitate the development of new technologies to surveil and threaten, and great technological spectacles to draw in allies. We've seen how the Mars of each of these periods was witnessed and understood through these understandings.

Over at least the past two centuries, the question of capitalism and the growth of markets, connected to concerns about the possible limits of resources, has dominated discussions of how we live in the world. In the late twentieth century, warnings about the impact of industrial practices on the environment and the need to regulate pollution and emissions became important elements of this discussion. And most recently, the idea of anthropogenic climate change—human-caused planetary-scale change—has caused us to consider whether life as we live it is a threat to life as we know it. On top of all this, we have begun to reckon with the global histories of colonialism, slavery, and genocide, and their impact on the world today.

We can certainly connect these questions to Mars. We have already seen how Mars can be imagined as a place for capitalism, labor, and resource extraction. But we can go further. The idea of the Anthropocene—that humans have initiated a geologic era in which they are the main force of change—can be read as a warning about the human potential for global destruction. But the Anthropocene has a Janus face. It can also be read as evidence of the ability of humans to change worlds. If on Earth this has happened by accident, and with unintended negative consequences, perhaps on Mars it can be directed in such a way that it creates a livable world. Elon Musk doesn't need to invoke science fiction ideas of terraforming from Arthur C. Clarke or Kim Stanley Robinson. He can tap into our present understanding of ourselves and our outsized impact on our own world when he suggests that we can change the atmosphere of Mars by dropping nuclear bombs on its surface. The Musk example also highlights that although the Cold War has ended, the weapons of the Cold War continue to threaten the world with another form of human-made destruction. As we've already seen in previous chapters, our exploration of Mars is facilitated by our past and present militarism.

We can read arguments for the colonization of Mars as reactions against historical accountability. Projecting a simplified and perhaps nostalgic image of the frontier past onto the future of Mars allows colonization proponents to celebrate American western expansion and Manifest Destiny on a landscape devoid of humans. This move continues the erasure that was central to the project of "settling" areas already populated by Indigenous peoples. It also continues the mythology of Eurocentrism we saw emerge in the sixteenth century: that Europeans and their descendants enact progress by moving into new lands and putting them to their proper use.

Right now, the idea of going to Mars does not dominate our culture. It mostly belongs to small communities of experts and enthusiasts. It has a larger cultural resonance, as it has for centuries, because it allows us to tell stories about ourselves that engage the imagination. This is as true about technical plans for Mars expeditions as it is about science fiction novels, films, and television series. If and when we do go to Mars, it will by necessity be such a massive undertaking that it will become one of the largest technological and cultural projects of its time. It will both shape and enact ideas about ourselves and our relationships to each other, our world, and the cosmos. For this reason, I think the most important question we can ask now is not "How will we get to Mars?" but "Who do we want to be when we become Martians?"

# ACKNOWLEDGMENTS

This book was written almost entirely during the COVID-19 pandemic of 2020–22. I could not have written the first few chapters without the assistance of the Smithsonian's libraries—in particular, our interlibrary loan staff and the participating institutions that scanned and provided chapters from books I couldn't otherwise access. I also benefited from the help of two contract historians who assisted me in gathering the sources used for this project: Matt Sanders and Peter Kleeman.

Some of what appears in the second half of the book is based on my experiences at Arizona State University with participants in the 2004 Mars Exploration Rover mission to Mars. I owe Phil Christensen a huge debt of gratitude for allowing me to tag along on his team's adventure, and I am very happy to be able to revisit that time in my life here, in this book. I also want to thank the historian of science Jane Maienschein, my advisor during those years, for helping me to make sense of what I was witnessing. Likewise, my PhD advisor, Naomi Oreskes, helped to shape my understanding of the Cold War transformation of earth and planetary science. Robert Westman and Luce Giard gave me an initial taste of the history of medieval and early modern science that until now I never had a chance to develop.

I was able to summarize the Mars chapter of Kircher's *Itinerarium* in chapter 3 thanks to the generosity of Ingrid Rowland and Jacqueline Glomski, who shared with me their own translations and notes on the text via email. Carlos Ziller Camenietzki was kind enough to share his Portuguese translation of Stansel's Latin text, from which I derived the English summary included here.

I wrote this book as a curator in the Space History Department of the Smithsonian's National Air and Space Museum. I want to thank the museum's director, Christopher Browne; the associate director of research and curatorial affairs, Jeremy Kinney; and the chair of the Space History Department, Margaret Weitekamp, for their support. I also wish to acknowledge Michael Neufeld and David DeVorkin for the feedback they provided on the first draft chapters as I finished them. Other colleagues, including Nicole Archambeau, Matthew Crawford, Rebecca Ljungren, Emily Margolis, Patrick McCray, and Jim Zimbelman, also provided valuable feedback on individual chapters. My colleagues in the Aeronautics and Space History Departments and the Center for Earth and Planetary Studies gave me excellent feedback when I presented some of this work at our semiweekly research seminar. Diego Jauregui and Mariana Zuniga, whom I supervised during their Smithsonian internships, provided feedback on ideas as I developed them.

Thank you to my editor, Karen Darling, for her support and encouragement in the writing of this book, and to her editorial assistant, Fabiola Enríquez, for helping me keep track of all the images and necessary permissions. And thank you to my manuscript editor, Renaldo Migaldi, for helping to untangle the flowers from the weeds in my prose.

# NOTES

**PROLOGUE**

1 Tony Greicius, "The Launch Is Approaching for NASA's Next Mars Rover, Perseverance," text, NASA, June 17, 2020; http://www.nasa.gov/feature/jpl/the-launch-is-approaching -for-nasas-next-mars-rover-perseverance.

2 Mike Wall, "NASA's next Mars Rover Carries Tribute to Healthcare Workers Fighting Coronavirus," Space.com, June 17, 2020; https://www.space.com/nasa-mars-rover -perseverance-coronavirus-tribute.html.

**INTRODUCTION**

1 Angel Tesorero, "Meet the Emirati Engineers of Hope Probe Mars Mission," *Gulf News*, February 10, 2021; https://gulfnews.com/special-reports/meet-the-emirati-engineers -of-hope-probe-mars-mission-1.77094280.

2 "The Space Review: Red Planet Scare," accessed May 25, 2021; https://thespacereview.com /article/4181/1?fbclid=IwAR158JQHovJ-LyMk6tcsx5QzeDZWgJqZBoNf9gPiirj2aG1x -BsfQc_cfzI.

3 Henry Neville Hutchinson, *The Autobiography of the Earth: A Popular Account of Geological History* (London: Edward Stanford, 1890), 2.

4 Kim Stanley Robinson, *Red Mars* (New York: Random House, 2003), 174.

**CHAPTER 1**

1 Thomas Hockey, *How We See the Sky: A Naked-Eye Tour of Day and Night* (Chicago: University of Chicago Press, 2011).

2 Steven Mithen, *The Prehistory of the Mind: A Search for the Origins of Art, Religion and Science* (London: Orion Books, 1998), 76.

3 Lawrence H. Robbins, "Astronomy and Prehistory," in *Astronomy across Cultures: The History of Non-Western Astronomy*, ed. Helaine Selin (Boston: Kluwer Academic Publishers, 2000), 37.

4  Anthony F. Aveni, *Ancient Astronomers* (Washington: Smithsonian, 1995), 23.

5  Aveni, *Ancient Astronomers*, 27.

6  Anthony F. Aveni, *Stairways to the Stars: Skywatching in Three Great Ancient Cultures* (New York: Wiley, 1997), 91.

7  Linda Schele and Mary Ellen Miller, *Blood of Kings: Dynasty and Ritual in Maya Art* (New York: George Braziller, 1986), 25.

8  Victoria R. Bricker and Harvey M. Bricker, "The Seasonal Table in the Dresden Codex and Related Almanacs," *Journal for the History of Astronomy Supplement* 19 (1988): S59.

9  William Sheehan and Jim Bell, *Discovering Mars: A History of Observation and Exploration of the Red Planet* (Tucson: University of Arizona Press, 2021), 3.

10  Sheehan and Bell, *Discovering Mars*, 7.

11  Harvey M. Bricker and Victoria R. Bricker, "More on the Mars Table in the Dresden Codex," *Latin American Antiquity* 8, no. 4 (1997): 389.

12  Anthony F. Aveni, Harvey M. Bricker, and Victoria R. Bricker, "Seeking the Sidereal: Observable Planetary Stations and the Ancient Maya Record," *Journal for the History of Astronomy* 34 (May 1, 2003): 159.

13  Harvey M. Bricker, Anthony F. Aveni, and Victoria R. Bricker, "Ancient Maya Documents Concerning the Movements of Mars," *Proceedings of the National Academy of Science* 98 (February 1, 2001): 2108.

14  Schele and Miller, *Blood of Kings*, 42.

15  Schele and Miller, *Blood of Kings*, 35.

16  Michael D. Coe and Mark Van Stone, *Reading the Maya Glyphs* (London: Thames & Hudson, 2005), 13.

17  David Freidel, Linda Schele, and Joy Parker, *Maya Cosmos: Three Thousand Years on the Shaman's Path* (New York: HarperCollins, 2001), 60.

18  Michael Bazzett, trans., *The Popol Vuh* (Minneapolis: Milkweed Editions, 2018), 6.

19  Freidel, Schele, and Parker, *Maya Cosmos*, 69.

20  Bazzett, *Popol Vuh*, 8.

21  Bazzett, *Popol Vuh*, 205.

22  Bazzett, *Popol Vuh*, 223.

23  Bazzett, *Popol Vuh*, 227.

24  Freidel, Schele, and Parker, *Maya Cosmos*, 67.

25  Freidel, Schele, and Parker, *Maya Cosmos*, 96.

26  Gerardo Aldana, *The Apotheosis of Jana ab' Pakal: Science, History, and Religion at Classic Maya Palenque* (Boulder: University Press of Colorado, 2010), 84.

27  Aldana, *Apotheosis of Janaab' Pakal*, 77.

28  Anthony F. Aveni, *Skywatchers: A Revised and Updated Version of Skywatchers of Ancient Mexico* (Austin: University of Texas Press, 2001), 169.

29  Aldana, *Apotheosis of Janaab' Pakal*, 190.

30  Liu An, *The Essential Huainanzi*, trans. John S. Major et al. (New York: Columbia University Press, 2012), 2.

31  Lillian Lan-ying Tseng, *Picturing Heaven in Early China* (Cambridge, MA: Harvard University Press, 2011), 17.

32  Sun Xiaochun, "Crossing the Boundaries between Heaven and Man: Astronomy in Ancient China," in *Astronomy across Cultures: The History of Non-Western Astronomy*, ed. Helaine Selin and Sun Xiaochun (Boston: Springer, 2000), 426.

33  Tseng, *Picturing Heaven in Early China*, 86.

34  Christopher Cullen, *Heavenly Numbers: Astronomy and Authority in Early Imperial China* (New York: Oxford University Press, 2017), 9.

35  Nathan Sivin, *Granting the Seasons: The Chinese Astronomical Reform of 1280, with a Study of Its Many Dimensions and a Translation of Its Records* (New York: Spring, 2009), 36.

36  Cullen, *Heavenly Numbers*, 47.

37  Quoted in Cullen, *Heavenly Numbers*, 47.

38  John S. Major, *Heaven and Earth in Early Han Thought: Chapters Three, Four, and Five of the Huainanzi* (Albany: State University of New York Press, 1993), 11.

39  Major, *Heaven and Earth in Early Han Thought*, 67.

40  Major, *Heaven and Earth in Early Han Thought*, 12.

41  An, *Essential Huainanzi*, 40.

42  Major, *Heaven and Earth in Early Han Thought*, 62.

43  Siven translates this as "dazzling delusion."

44  Major, *Heaven and Earth in Early Han Thought*, 71.

45  Major, *Heaven and Earth in Early Han Thought*, 74.

46  Nicholas Campion, *A History of Western Astrology, Volume 1: The Ancient World* (New York: Bloomsbury Academic, 2008), 36.

47  Francesca Rochberg, *The Heavenly Writing: Divination, Horoscopy, and Astronomy in Mesopotamian Culture* (New York: Cambridge University Press, 2004), 36.

48  Francesca Rochberg, *In the Path of the Moon: Babylonian Celestial Divination and Its Legacy* (Boston: Brill, 2010), 34.

49  Rochberg, *Heavenly Writing*, 32.

50  Noel M. Swerdlow, *The Babylonian Theory of the Planets* (Princeton, NJ: Princeton University Press, 1998), 5.

51  Campion, *History of Western Astrology*, 75.

52  Swerdlow, *Babylonian Theory of the Planets*, 33.

53  W. G. Lambert's translation of the *Enuma Elish* can be found online: W. G. Lambert, "Enuma Elish: The Babylonian Epic of Creation," *World History Encyclopedia*, accessed April 10, 2021; https://www.ancient.eu/article/225/enuma-elish---the-babylonian-epic-of-creation---fu/.

54  Lambert "*Enuma Elish.*"

55  Rochberg, *Heavenly Writing*, 59.

56  Campion, *History of Western Astrology*, 54.

57  Rochberg, *Heavenly Writing*, 65; Swerdlow, *Babylonian Theory of the Planets*, 23.

58  Campion, *History of Western Astrology*, 56.

59  Swerdlow, *Babylonian Theory of the Planets*, 16.

60  Swerdlow, *Babylonian Theory of the Planets*, 9.

61  Swerdlow, *Babylonian Theory of the Planets*, 12.

62  Swerdlow, *Babylonian Theory of the Planets*, 12.

63  Rochberg, *Heavenly Writing*, 60.

64  Rochberg, *Heavenly Writing*, 76.

65  David C. Lindberg, *The Beginnings of Western Science: The European Scientific Tradition in Philosophical, Religious, and Institutional Context, Prehistory to A.D. 1450* (Chicago: University of Chicago Press, 2008), 27.

66  Lindberg, *Beginnings of Western Science*, 40.

67  Lindberg, *Beginnings of Western Science*, 43.

68  Lindberg, *Beginnings of Western Science*, 54.

69  Lindberg, *Beginnings of Western Science*, 62.

70  Lindberg, *Beginnings of Western Science*, 99.

71  Lindberg, *Beginnings of Western Science*, 104.

### CHAPTER 2

1  The Paris Concilium is translated in Rosemary Horrox, *The Black Death* (Manchester, UK: Manchester University Press, 1994), 159.

2  Horrox, *Black Death*, 161.

3  Genevieve Miller, "'Airs, Waters, and Places' in History," *Journal of the History of Medicine and Allied Sciences* 17, no. 1 (January 1, 1962): 137.

4  De Maux's treatise on the Plague is translated in Horrox, *Black Death*, 168.

5  Edward Grant, "Cosmology," in *The Cambridge History of Science*, 2013, 436.

6  Edward Grant, "The Medieval Cosmos: Its Structure and Operation," *Journal for the History of Astronomy* 28, no. 2 (May 1, 1997): 162.

7  Edward Grant, *Planets, Stars, and Orbs: The Medieval Cosmos, 1200–1687* (New York: Cambridge University Press, 1994), 588.

8  Grant, *Planets, Stars, and Orbs*, 222, 424.

9  Grant, *Planets, Stars, and Orbs*, 587.

10  Grant, *Planets, Stars, and Orbs*, 587.

11  Claudius Ptolemaeus, *Tetrabiblos*, trans. Frank E. Robbins (Cambridge, MA: Harvard University Press, 1940), 37.

12  Ptolemaeus, *Tetrabiblos*, 41.

13  Ptolemaeus, *Tetrabiblos*, 183.

14  Grant, *Planets, Stars, and Orbs*, 204.

15  Joseph Patrick Byrne, *Daily Life during the Black Death* (Westport, CT: Greenwood Publishing Group, 2006), 16.

16  Justine Isserles, "Bloodletting and Medical Astrology in Hebrew Manuscripts from Medieval Western Europe," *Sudhoffs Archiv* 101, no. 1 (2017): 19.

17  Markham Judah Geller, *Melothesia in Babylonia: Medicine, Magic, and Astrology in the Ancient Near East* (Boston: Walter de Gruyter, 2014), 88.

18  Janet L. Abu-Lughod, *Before European Hegemony: The World System A.D. 1250–1350* (New York: Oxford University Press, 1991), 21.

19  Lindberg, *Beginnings of Western Science*, 153.

20  Edward Grant, *The Foundations of Modern Science in the Middle Ages: Their Religious, Institutional and Intellectual Contexts* (New York: Cambridge University Press, 1996), 13.

21  Joel L. Kraemer, *Humanism in the Renaissance of Islam: The Cultural Revival during the Buyid Age* (Leiden, Netherlands: E. J. Brill, 1986), 12.

22  A. I. Sabra, "Situating Arabic Science: Locality versus Essence," *Isis* 87, no. 4 (1996): 658; Dimitri Gutas, *Greek Thought, Arabic Culture: The Graeco-Arabic Translation Movement in Baghdad and Early 'Abbasaid Society (2nd–4th/5th–10th c.)* (London: Routledge, 2012), 34.

23  A. I. Sabra, "The Appropriation and Subsequent Naturalization of Greek Science in Medieval Islam: A Preliminary Statement," *History of Science* 25, no. 3 (September 1, 1987): 228; Bernard R. Goldstein, "Astronomy and the Jewish Community in Early Islam," *Aleph*, no. 1 (2001): 18.

24  Bernard R. Goldstein, "Astronomy as a 'Neutral Zone': Interreligious Cooperation In Medieval Spain," *Medieval Encounters* 15, no. 2 (2009): 161.

25  Bernard R. Goldstein, "The Making of Astronomy in Early Islam," *Nuncius* 1, no. 2 (January 1, 1986): 80.

26  Goldstein, "Making of Astronomy," 82.

27  Goldstein, "Making of Astronomy," 87.

28  Goldstein, "Making of Astronomy," 87.

29  Abu Ma'shar, *The Abbreviation of the Introduction to Astrology: Together with the Medieval Latin Translation of Adelard of Bath*, trans. Charles Burnett, Keiji Yamamoto and Michio Yano (New York: E. J. Brill, 1994).

30  Robert G. Morrison, "Islamic Astronomy," in *The Cambridge History of Science, Volume 2: Medieval Science*, ed. David C. Lindberg and Michael H. Shank (Cambridge, UK: Cambridge University Press, 2013), 126.

31  Morrison, "Islamic Astronomy," 114.

32  Abu-Lughod, *Before European Hegemony*, 47.

33  Grant, "Cosmology," 436.

34  See, for example, Richard Kieckhefer, *Forbidden Rites: A Necromancer's Manual of the Fifteenth Century* (Stroud, UK: Sutton, 1997).

35  Juris Lidaka, "The Book of Angels, Rings, Characters, and Images of the Planets: Attributed to Osbern Bokenham," in *Conjuring Spirits: Texts and Traditions of Late Medieval Ritual Magic*, ed. Claire Fanger (University Park: Pennsylvania State University Press, 1994), 67.

36  Lidaka, "Book of Angels," 49.

37  Kieckhefer, *Forbidden Rites*, 75.

38  Lidaka, "Book of Angels," 51.

39  Mark A. Waddell, *Magic, Science, and Religion in Early Modern Europe*, New Approaches to the History of Science and Medicine (New York: Cambridge University Press, 2021), 15.

40  Michael A. Ryan, *A Kingdom of Stargazers: Astrology and Authority in the Late Medieval Crown of Aragon* (Cornell University Press, 2011), 16.

41  Grant, *Planets, Stars, and Orbs*, 426.

42  Mark Kauntze, *Authority and Imitation: A Study of the* Cosmographia *of Bernard Silvestris* (Boston: Brill, 2014), 54.

43  Martianus Capella, *Martianus Capella and the Seven Liberal Arts: The Marriage of Philology and Mercury*, trans. William Harris Stahl (New York: Columbia University Press, 1971), 63.

44  Capella, *Martianus Capella and the Seven Liberal Arts*, 51.

45  Capella, *Martianus Capella and the Seven Liberal Arts*, 59.

46  Kauntze, *Authority and Imitation*, 79.

47  Ambrosius Aurelius Theodosius Macrobius, *Commentary on the Dream of Scipio*, trans. William Harris Stahl (New York: Columbia University Press, 1990), 130.

48  Macrobius, *Commentary*, 137.

49  Macrobius, *Commentary*, 136.

50  Anicius Manlius Severinus Boethius, *The Consolation of Philosophy*, trans. Richard H. Green (Mineola, NY: Dover Publications, 2002), 68.

51  Kauntze, *Authority and Imitation*, 60.

52 Bernardus Silvestris, *The Cosmographia of Bernardus Silvestris*, trans. Winthrop Wetherbee (New York: Columbia University Press, 1990), 75.

53 Silvestris, *Cosmographia*, 78.

54 Silvestris, *Cosmographia*, 89.

55 Kauntze, *Authority and Imitation*, 65.

56 Silvestris, *Cosmographia*, 98.

57 Silvestris, *Cosmographia*, 101.

58 Kauntze, *Authority and Imitation*, 69.

59 Kauntze, *Authority and Imitation*, 83.

60 Kauntze, *Authority and Imitation*, 89.

61 Miguel Asín y Palacios, *Islam and the Divine Comedy*, trans. Harold Sunderland (London: John Murray, 1926).

62 Dante Alighieri, *Il Convito*, trans. Elizabeth Price Sayer (New York: G. Routledge and Sons, 1887), 86.

### CHAPTER 3

1 John M. Hobson, *The Eastern Origins of Western Civilisation* (New York: Cambridge University Press, 2004), 116.

2 Elizabeth Horodowich, *The Venetian Discovery of America: Geographic Imagination and Print Culture in the Age of Encounters* (New York: Cambridge University Press, 2018), 15.

3 Evidence of the global Scientific Revolution has been accumulating steadily, and was recently synthesized in James Poskett, *Horizons: The Global Origins of Modern Science* (Boston: Mariner Books, 2022).

4 Historians believe that Martin Behaim built two such globes. The one that survives is held at the Germanisches Nationalmuseum in Nuremberg, Germany.

5 Eugene F. Rice and Anthony Grafton, *The Foundations of Early Modern Europe, 1460–1559*, 2nd ed. (New York: W. W. Norton, 1994), 7.

6 Anthony Grafton, April Shelford, and Nancy Siraisi, *New Worlds, Ancient Texts: The Power of Tradition and the Shock of Discovery* (Cambridge, MA: Harvard University Press, 1995), 22.

7 Elly Dekker, "Globes in Renaissance Europe," in *The History of Cartography, Volume 3*, ed. David Woodward (Chicago: University of Chicago Press, 2007), 138.

8 Ptolemaeus, *Tetrabiblos*, 125.

9 Ptolemaeus, *Tetrabiblos*, 127.

10 Ptolemaeus, *Tetrabiblos*, 137.

11 Nicolás Wey Gómez, *The Tropics of Empire: Why Columbus Sailed South to the Indies* (Cambridge, MA: MIT Press, 2008), 79.

12 Gómez, *Tropics of Empire*, 66.

13 Gómez, *Tropics of Empire*, 61.

14 Grant, *Planets, Stars, and Orbs*, 452.

15 Gómez, *Tropics of Empire*, 11.

16 Benson Bobrick, *The Fated Sky: Astrology in History* (New York: Simon & Schuster, 2006), 5; George E. Nunn, "The Imago Mundi and Columbus," *American Historical Review* 40, no. 4 (July 1, 1935): 646–61; Grafton, Shelford, and Siraisi, *New Worlds, Ancient Texts*, 77.

17 Acosta, quoted in Grafton, Shelford, and Siraisi, *New Worlds, Ancient Texts*, 1.

18  Grafton, Shelford, and Siraisi, *New Worlds, Ancient Texts*, 115.

19  Robert Westman, *The Copernican Question: Prognostication, Skepticism, and Celestial Order* (Berkeley: University of California Press, 2011).

20  Antonio Barrera-Osorio, *Experiencing Nature: The Spanish American Empire and the Early Scientific Revolution* (Austin: University of Texas Press, 2010).

21  Peter Dear, *Revolutionizing the Sciences: European Knowledge in Transition, 1500–1700*, 3rd ed. (London: Red Globe Press, 2019), 170.

22  Harald Siebert, "The Early Search for Stellar Parallax: Galileo, Castelli, and Ramponi," *Journal for the History of Astronomy* 36 (August 1, 2005): 253.

23  John L. Heilbron, *Galileo* (Oxford, UK: Oxford University Press, 2012), 151.

24  Mary Baine Campbell, *Wonder and Science: Imagining Worlds in Early Modern Europe* (Ithaca, NY: Cornell University Press, 2004), 127.

25  Campbell, *Wonder and Science*, 130; Westman, *Copernican Question*, 455.

26  Horodowich, *Venetian Discovery of America*, 20.

27  Praise from Galileo's contemporaries quoted in Heilbron, *Galileo*, 165.

28  Elizabeth Horodowich, "Italy and the New World," in *The New World in Early Modern Italy, 1492–1750*, ed. Elizabeth Horodowich and Lia Markey (New York: Cambridge University Press, 2018), 22.

29  Martin Kemp, "Moving in Elevated Circles," *Nature* 466, no. 7302 (July 2010): 33.

30  Mark Rosen, "A New Chronology of the Construction and Restoration of the Medici Guardaroba in the Palazzo Vecchio, Florence," *Mitteilungen des Kunsthistorischen Institutes in Florenz* 53, no. 2/3 (2009): 285–308.

31  Lia Markey, *Imagining the Americas in Medici Florence* (University Park: Pennsylvania State University Press, 2016), 2.

32  Ingrid D. Rowland, "Athanasius Kircher, Giordano Bruno, and the Panspermia of the Infinite Universe," in *Athanasius Kircher: The Last Man Who Knew Everything*, ed. Paula Findlen (New York: Routledge, 2004), 192.

33  Jacqueline Glomski, "Religion, the Cosmos, and Counter-Reformation Latin: Athanasius Kircher's Itinerarium Exstaticum (1656)," *Acta Conventus Neo-Latini Monasteriensis: Proceedings of the Fifteenth International Congress of Neo-Latin Studies (Münster 2012)* (March 2015): 228.

34  Ingrid D. Rowland, "Poetry and Prophecy in the Encyclopedic System of Athanasius Kircher," *Bruniana & Campanelliana* 11, no. 2 (2005): 511.

35  Ingrid Rowland, "'Th' United Sense of th' Universe': Athanasius Kircher in Piazza Navona," *Memoirs of the American Academy in Rome* 46 (2001): 153; Ingrid D. Rowland, *The Ecstatic Journey: Athanasius Kircher in Baroque Rome* (Chicago: University of Chicago Library, 2000).

36  Daniel Stolzenberg, *Egyptian Oedipus: Athanasius Kircher and the Secrets of Antiquity* (Chicago: University of Chicago Press, 2013), 25.

37  John Edward Fletcher, *A Study of the Life and Works of Athanasius Kircher, "Germanus Incredibilis"* (Boston: Brill, 2011), 136.

38  From Kircher's Mundus Subterraneus, quoted in Rowland, "Th' United Sense of th' Universe," 169–70.

39  Quoted in Fletcher, *Life and Works of Athanasius Kircher*, 139.

40  Rowland, "Poetry and Prophecy," 314.

41  Glomski, "Religion, the Cosmos, and Counter-Reformation Latin," 232.

42  Glomski, "Religion, the Cosmos, and Counter-Reformation Latin," 233.

43  Rowland, "Athanasius Kircher, Giordano Bruno, and the Panspermia of the Infinite Universe," 193.

44  Rowland, "Athanasius Kircher, Giordano Bruno, and the Panspermia of the Infinite Universe," 194.

45  Carlos Ziller Camenietzki, "Baroque Science between the Old and the New World: Father Kircher and His Colleague Valentin Stansel (1621–1705)," in *Athanasius Kircher: The Last Man Who Knew Everything*, ed. Paula Findlen (New York: Routledge, 2004), 317.

46  Lawrence M. Principe, *The Secrets of Alchemy*, illustrated edition (Chicago: University of Chicago Press, 2015), 13.

47  Principe, *Secrets of Alchemy*, 205.

48  Camenietzki, "Baroque Science," 321.

49  There is no published English translation of Kircher's *Itinerarium exstaticum*. I am able to summarize the Mars chapter of the *Itinerarium* here thanks to the generosity of Ingrid Rowland and Jacqueline Glomski, who shared their own translations and notes with me. This passage is translated by Rowland.

50  Glomski, "Religion, the Cosmos, and Counter-Reformation Latin."

51  From Rowland's translation.

52  Glomski, "Religion, the Cosmos, and Counter-Reformation Latin."

53  No English translation of Stansel's *Uranophilus caelestis peregrinus* currently exists. Carlos Ziller Camenietzki was kind enough to share his Portuguese translation of the Latin text, from which I have produced the English summary included here.

54  Westman, *Copernican Question*, 510.

55  Maurice A. Finocchiaro, ed., *The Galileo Affair: A Documentary History* (Berkeley: University of California Press, 1989), 65.

56  Sheehan and Bell, *Discovering Mars*, 39.

57  Richard S. Westfall, *The Construction of Modern Science: Mechanisms and Mechanics* (New York: Cambridge University Press, 1977), 29.

58  John Hedley Brooke, *Science and Religion: Some Historical Perspectives* (New York: Cambridge University Press, 1991), 119.

59  Barrera-Osorio, *Experiencing Nature*.

60  Bacon, quoted in Grafton, Shelford, and Siraisi, *New Worlds, Ancient Texts*, 198.

### CHAPTER 4

1  Westfall, *Construction of Modern Science*, 121.

2  Steven J. Dick, *The Biological Universe: The Twentieth Century Extraterrestrial Life Debate and the Limits of Science* (New York: Cambridge University Press, 1996), 18.

3  Robert Markley, *Dying Planet: Mars in Science and the Imagination* (Durham, NC: Duke University Press, 2005), 41.

4  Westfall, *Construction of Modern Science*, 143.

5  John Henry, "Primary and Secondary Causation in Samuel Clarke's and Isaac Newton's Theories of Gravity," *Isis* 111, no. 3 (September 1, 2020): 558.

6  Brooke, *Science and Religion*, 139.

7  Richard S. Westfall, *The Life of Isaac Newton* (New York: Cambridge University Press, 1993), 116.

8  Henry, "Primary and Secondary Causation," 550.

9   Quoted in Henry, "Primary and Secondary Causation," 556.

10  Brooke, *Science and Religion*, 146.

11  Ronald L. Numbers, *Creation by Natural Law: Laplace's Nebular Hypothesis in American Thought* (Seattle: University of Washington Press, 1977), 4.

12  Stephen G. Brush, "The Nebular Hypothesis and the Evolutionary Worldview," *History of Science* 25, no. 3 (September 1, 1987): 245.

13  Roger Hahn, "Laplace and the Mechanical Universe," in *God and Nature: Historical Essays on the Encounter between Christianity and Science*, ed. David C. Lindberg and Ronald L. Numbers (Berkeley: University of California Press, 1986), 257.

14  Numbers, *Creation by Natural Law*, 3.

15  Kenneth L. Taylor, "Earth and Heaven, 1750–1800: Enlightenment Ideas about the Relevance to Geology of Extraterrestrial Operations and Events," *Earth Sciences History* 17, no. 2 (1998): 84.

16  There is no translation for *Telliamed*, as it is de Maillet's name spelled backwards.

17  Taylor, "Earth and Heaven, 1750–1800," 85.

18  Taylor, "Earth and Heaven, 1750–1800," 85.

19  Numbers, *Creation by Natural Law*, 6.

20  Both Maillet and Buffon believed that the marine fossils found in mountain ranges were empirical evidence of a former global ocean and the gradual drying of Earth; Peter J. Bowler, *Evolution: The History of an Idea* (Berkeley: University of California Press, 2009), 59.

21  Brush, "Nebular Hypothesis," 260.

22  Simon Schaffer, "Herschel in Bedlam: Natural History and Stellar Astronomy," *British Journal for the History of Science* 13, no. 3 (1980): 211.

23  Schaffer, "Herschel in Bedlam," 227.

24  Simon Schaffer, "'The Great Laboratories of the Universe': William Herschel on Matter Theory and Planetary Life," *Journal for the History of Astronomy* 11 (1980): 99.

25  Herschel quoted in Markley, *Dying Planet*, 48.

26  Bernard Lightman, *Victorian Popularizers of Science: Designing Nature for New Audiences* (Chicago: University of Chicago Press, 2007), 221.

27  James A. Secord, *Victorian Sensation: The Extraordinary Publication, Reception, and Secret Authorship of Vestiges of the Natural History of Creation* (Chicago: University of Chicago Press, 2000), 98.

28  Secord, *Victorian Sensation*, 87; Marilyn Bailey Ogilvie, "Robert Chambers and the Nebular Hypothesis," *British Journal for the History of Science* 8, no. 3 (1975): 214.

29  Robert Chambers, *Vestiges of the Natural History of Creation and Other Evolutionary Writings*, ed. James A. Secord (Chicago: University of Chicago Press, 1994), 307.

30  Chambers, *Vestiges*, 12.

31  Bernard Lightman, "'The Voices of Nature': Popularizing Victorian Science," in *Victorian Science in Context* (Chicago: University of Chicago Press, 1997), 200.

32  Markley, *Dying Planet*, 50.

33  Joshua Nall, *News from Mars: Mass Media and the Forging of a New Astronomy, 1860–1910* (Pittsburgh: University of Pittsburgh Press, 2019), 9.

34  Proctor quoted in Bernard Lightman, "The Visual Theology of Victorian Popularizers of Science: From Reverent Eye to Chemical Retina," *Isis* 91, no. 4 (2000): 670.

35  Proctor quoted in Nall, *News from Mars*, 34.

36  Schiaparelli quoted in K. Maria D. Lane, "Geographers of Mars: Cartographic Inscription and Exploration Narrative in Late Victorian Representations of the Red Planet," *Isis* 96, no. 4 (2005): 494.

37  Schiaparelli quoted in Dick, *Biological Universe*, 69.

38  Schiaparelli quoted in Dick, *Biological Universe*, 69.

39  Schiaparelli quoted in Dick, *Biological Universe*, 70.

40  Dick, *Biological Universe*, 70.

41  Martin Willis, *Vision, Science and Literature, 1870–1920: Ocular Horizons* (Pittsburgh: University of Pittsburgh Press, 2016), 61.

42  K. Maria D. Lane, "Mapping the Mars Canal Mania: Cartographic Projection and the Creation of a Popular Icon," *Imago Mundi* 58, no. 2 (2006): 201.

43  Lane, "Geographers of Mars," 478.

44  Lane, "Geographers of Mars," 490.

45  Nall, *News from Mars*, 89.

46  Nall, *News from Mars*, 107.

47  Nall, *News from Mars*, 111.

48  Nall, *News from Mars*, 122.

49  Pickering quoted in Nall, *News from Mars*, 117.

50  Lane, "Geographers of Mars," 498.

51  Dick, *Biological Universe*, 71.

52  Dick, *Biological Universe*, 74.

53  Nall, *News from Mars*, 133.

54  Dick, *Biological Universe*, 76.

55  Lowell quoted in William Graves Hoyt, *Lowell and Mars* (Tucson: University of Arizona Press, 1996), 80.

56  Markley, *Dying Planet*, 67; Dick, *Biological Universe*, 77.

57  Rainer Eisfeld, "Projecting Landscapes of the Human Mind onto Another World: Changing Faces of an Imaginary Mars," in *Radical Approaches to Political Science: Roads Less Traveled* (Berlin: Barbara Budrich, 2012), 191.

58  Lane, "Geographers of Mars," 502.

59  Poem quoted in Hoyt, *Lowell and Mars*, 221.

60  Markley, *Dying Planet*, 53.

61  Lowell (1909) quoted in Markley, *Dying Planet*, 68.

62  Antoniadi quoted in Dick, *Biological Universe*, 93.

63  Antoniadi to Lowell, quoted in Hoyt, *Lowell and Mars*, 169.

64  Antoniadi quoted in Lane, "Mapping the Mars Canal Mania," 207.

65  Antoniadi quoted in Dick, *Biological Universe*, 99.

66  Dick, *Biological Universe*, 95.

67  Karl S. Guthke, *Exploring the Interior: Essays on Literary and Cultural History* (Cambridge, UK: Open Book Publishers, 2018), 7.

68  Guthke, *Exploring the Interior*, 192.

69  Guthke, *Exploring the Interior*, 201.

70  Miles Wilson, *The History of Israel Jobson, the Wandering Jew* (London: J. Nicholson, 1757), 30.

71  Karl S. Guthke, *The Last Frontier: Imagining Other Worlds from the Copernican Revolution to Modern Science Fiction* (Ithaca, NY: Cornell University Press, 1990), 367.

72  Percy Greg, *Across the Zodiac: The Story of a Wrecked Record* (London: Trübner, 1880).

73  Willis, *Vision, Science and Literature*, 59.

74  Willis, *Vision, Science and Literature*, 82.

75  Robert Crossley, *Imagining Mars: A Literary History* (Middletown, CT: Wesleyan University Press, 2011), 90.

76  Crossley, *Imagining Mars*, 118.

77  Eisfeld, "Projecting Landscapes of the Human Mind," 192.

78  Markley, *Dying Planet*, 118.

79  Sam Moskowitz, ed., *Under the Moons of Mars: A History and Anthology of the Scientific Romance in the Munsey Magazines, 1912–1920* (New York: Holt Reinhart & Winston, 1970), 291.

80  Crossley, *Imagining Mars*, 153.

### CHAPTER 5

1  Naomi Oreskes, *Science on a Mission: How Military Funding Shaped What We Do and Don't Know about the Ocean* (Chicago: University of Chicago Press, 2021), 4.

2  Chandra Mukerji, *A Fragile Power: Scientists and the State* (Princeton, NJ: Princeton University Press, 1989), 5.

3  Peter J. Westwick, *Into the Black: JPL and the American Space Program, 1976–2004* (New Haven, CT: Yale University Press, 2007), 98.

4  David DeVorkin, *Science with a Vengeance: How the Military Created the US Space Sciences after World War II* (New York: Springer-Verlag, 1992), 342.

5  Edward Clinton Ezell and Linda Neuman Ezell, *On Mars: Exploration of the Red Planet, 1958–1978* (Washington: National Aeronautics and Space Administration, 1984), 24.

6  Westwick, *Into the Black*, 4.

7  From Pickering's foreword to Jet Propulsion Laboratory (US), *Mariner Mission to Venus: Prepared for the National Aeronautics and Space Administration* (New York: McGraw-Hill, 1963).

8  Jet Propulsion Laboratory (US), *Mariner Mission to Venus*, 2.

9  D. Briot, J. Schneider, and L. Arnold, "G. A. Tikhov, and the Beginnings of Astrobiology," *Extrasolar Planets: Today and Tomorrow* 321 (December 1, 2004): 219–20.

10  William M. Sinton, "Further Evidence of Vegetation on Mars," *Science* 130, no. 3384 (1959): 1234–37.

11  Jordan Bimm, "Green Mars: Mars Jars and The Origin of Strughold's Vegetation Hypothesis," *Hedgehog*, fall 2020: 4-7.

12  Albert G. Wilson et al., "Problems Common to the Fields of Astronomy and Biology: A Symposium," *Publications of the Astronomical Society of the Pacific* 70, no. 412 (1958): 45.

13  Ezell and Ezell, *On Mars*, 77.

14  Ezell and Ezell, *On Mars*, 79.

15  W. Henry Lambright, *Why Mars: NASA and the Politics of Space Exploration* (Baltimore: Johns Hopkins University Press, 2014), 35.

16  Clayton R. Koppes, *JPL and the American Space Program: A History of the Jet Propulsion Laboratory* (New Haven, CT: Yale University Press, 1982), 185.

17  Koppes, *JPL and the American Space Program*, 196.

18  Koppes, *JPL and the American Space Program*, 202.

19  Lambright, *Why Mars*.

20  Matthew Shindell, "Geophysics," in *A Companion to the History of American Science* (Boston: John Wiley & Sons, 2015), 130; Matthew Shindell, "Domesticating the Planets: Instruments and Practices in the Development of Planetary Geology," *Spontaneous Generations: A Journal for the History and Philosophy of Science* 4, no. 1 (2010): 191–230; Matthew Shindell, "From the End of the World to the Age of the Earth: The Cold War Development of Isotope Geochemistry at the University of Chicago and Caltech," in *Science and Technology in the Global Cold War*, ed. Naomi Oreskes and John Krige (Cambridge: MIT Press, 2014).

21  Westwick, *Into the Black*, 98.

22  Bimm, "Green Mars."

23  W. Henry Lambright, "Big Science in Space: Viking, Cassini, and the Hubble Space Telescope," in *Exploring the Solar System: The History and Science of Planetary Exploration*, ed. Roger D. Launius (New York: Palgrave Macmillan, 2013), 130.

24  Lambright, "Big Science in Space," 131.

25  Lambright, *Why Mars*, 38.

26  Agnew quoted in John M. Logsdon, *After Apollo? Richard Nixon and the American Space Program* (New York: Palgrave Macmillan, 2015), 64.

27  Michael Neufeld, *Von Braun: Dreamer of Space, Engineer of War* (New York: Vintage Books, 2008), 240.

28  Neufeld, *Von Braun*, 243.

29  Neufeld, *Von Braun*, 241.

30  Space Task Group, "The Post-Apollo Space Program: Directions for the Future," September 1969; available in NASA Historical Reference Collection, History Office, NASA Headquarters, Washington.

31  Paine quoted in Logsdon, *After Apollo?* 59.

32  Lambright, "Big Science in Space," 132.

33  Koppes, *JPL and the American Space Program*, 219.

34  Wesley T. Huntress Jr. and Mikhail Ya. Marov, *Soviet Robots in the Solar System: Mission Technologies and Discoveries* (New York: Springer, 2011), 259.

35  *Mars as Viewed by Mariner 9: A Pictorial Presentation by the Mariner 9 Television Team and the Planetology Program Principal Investigators* (Washington: National Aeronautics and Space Administration, 1974), 1.

36  *Mars as Viewed by Mariner 9*, preface.

37  Quoted in Koppes, *JPL and the American Space Program*, 219.

38  Thomas A. Mutch et al., *The Geology of Mars* (Princeton, NJ: Princeton University Press, 1976), 38.

39  Michael H. Carr, *The Surface of Mars* (New Haven, CT: Yale University Press, 1981), 135.

40  *Mars as Viewed by Mariner 9*, 41.

41  *Mars as Viewed by Mariner 9*, 185.

42  Koppes, *JPL and the American Space Program*, 220.

43  Robert Markley, "Missions to Mars: Reimagining the Red Planet in the Age of Spaceflight," in *Exploring the Solar System: The History and Science of Planetary Exploration*, ed. Roger D. Launius (New York: Palgrave Macmillan, 2013), 251.

44  Lambright, "Big Science in Space," 134.

45  James Tomayko, *Computers in Spaceflight: The NASA Experience* (Washington: National Aeronautics and Space Administration, 1988), 163.

46  Ezell and Ezell, *On Mars*, 408.

47  Lambright, "Big Science in Space," 134.

48  Kenneth L. Tanaka, "The Stratigraphy of Mars," *Journal of Geophysical Research* 91, no. B13 (November 30, 1986): E139–58.

49  Carr, *Surface of Mars*, 203.

50  Carr, *Surface of Mars*, 204.

51  Carr, *Surface of Mars*, 205.

52  Bruce Murray, Michael C. Malin, and Ronald Greeley, *Earthlike Planets: Surfaces of Mercury, Venus, Earth, Moon, Mars* (San Francisco: W. H. Freeman, 1981), 322.

53  Murray, Malin, and Greeley, *Earthlike Planets*, 345.

54  Murray, Malin, and Greeley, *Earthlike Planets*, 346.

55  De Witt Douglas Kilgore, *Astrofuturism: Science, Race, and Visions of Utopia in Space* (Philadelphia: University of Pennsylvania Press, 2003), 38.

56  W. Patrick McCray, *The Visioneers: How a Group of Elite Scientists Pursued Space Colonies, Nanotechnologies, and a Limitless Future* (Princeton, NJ: Princeton University Press, 2013).

57  Carl Sagan, *Cosmos* (New York: Ballantine Books, 2013), 138.

58  Markley, *Dying Planet*, 295.

59  John Grant, "Gaining Inspiration from the Martian Chronicles," *Stories from the Smithsonian National Air and Space Museum* (blog), August 20, 2020, https://airandspace.si.edu /stories/editorial/gaining-inspiration-martian-chronicles.

60  Thomas J. Morrissey, "Ready or Not, Here We Come: Metaphors of the Martian Megatext from Wells to Robinson," *Journal of the Fantastic in the Arts* 10, no. 4 (40) (2000): 386.

61  Philip K. Dick, "We Can Remember It for You Wholesale," in *The Wesleyan Anthology of Science Fiction*, ed. Arthur B. Evans et al. (Middletown, CT: Wesleyan University Press, 2010), 393.

62  Paul Roberts, "Ego Trip," *Cinefex* 43 (August 1990): 17.

63  Eric Brevig, interview by Matthew Shindell, June 18, 2021.

64  Eric Brevig to Matthew Shindell, "Total Recall Miniature Sets," July 7, 2021.

65  David Pieri, interview by Matthew Shindell, July 10, 2021.

66  Mark Stetson, interview by Matthew Shindell, July 11, 2021.

67  Robert C. Kiviat, "Casting a New Light on the Mars Face," *Omni* 16, no. 11 (August 1, 1994): 31–37.

## CHAPTER 6

1  Jacob Margolis, "How a Tweet about the Mars Rover Dying Blew Up on the Internet and Made People Cry," *LAist*, February 16, 2019, https://laist.com/news/jpl-mars-rover -opportunity-battery-is-low-and-its-getting-dark.

2  Sarah Stewart Johnson, *The Sirens of Mars: Searching for Life on Another World* (New York: Crown, 2020), 144.

3  Jason Filiatrault, "Goodbye, Opportunity Rover," *Washington Post*, February 13, 2019; https://www.washingtonpost.com/opinions/2019/02/13/goodbye-opportunity-rover -thank-you-letting-humanity-see-mars-with-your-eyes/.

4  Burkitt, "Future of Spirit," *Making XKCD Slightly Worse*, January 4, 2014; http://xkcdsw .com/3968.

5  Francis Spufford, *I May Be Some Time: Ice and the English Imagination* (New York: St. Martin's Press, 1997).

6 Flora Lichtman and Sharon Shattuck, "Animated Life: Pangea," *New York Times*, February 17, 2015; https://www.nytimes.com/video/opinion/100000003515124/animated-life-pangaea.html.

7 Lambright, *Why Mars*, 259.

8 Pilcher quoted in Erik M. Conway, *Exploration and Engineering: The Jet Propulsion Laboratory and the Quest for Mars* (Baltimore: Johns Hopkins University Press, 2015), 54.

9 Lambright, *Why Mars*. For a political history of JPL and how its management responded to these shifting political winds, see Westwick, *Into the Black*. For a detailed account of how these changes affected engineering practice and projects at JPL, see Conway, *Exploration and Engineering*.

10 William J. Clancey, *Working on Mars: Voyages of Scientific Discovery with the Mars Exploration Rovers* (Cambridge, MA: MIT Press, 2012).

11 Janet Vertesi, *Seeing Like a Rover: How Robots, Teams, and Images Craft Knowledge of Mars* (Chicago: University of Chicago Press, 2015).

12 Lisa Messeri, *Placing Outer Space: An Earthly Ethnography of Other Worlds* (Durham, NC: Duke University Press, 2016).

13 Zara Mirmalek, *Making Time on Mars* (Cambridge, MA: MIT Press, 2020).

14 Westwick, *Into the Black*, 178.

15 Conway, *Exploration and Engineering*, 31.

16 Shindell, "Domesticating the Planets."

17 Kathy Sawyer, "A Mars Never Dreamed Of," *National Geographic*, February 2001; https://www.nationalgeographic.com/science/article/mars-explored-anew.

18 P. R. Christensen et al., "Detection of Crystalline Hematite Mineralization on Mars by the Thermal Emission Spectrometer: Evidence for Near-Surface Water," *Journal of Geophysical Research: Planets* 105, no. E4 (2000): 9623–42; Philip R. Christensen and Matthew B. Shindell, "Mars Infrared," *Planetary Report*, June 2003.

19 Michael J. Neufeld, "Transforming Solar System Exploration: The Origins of the Discovery Program, 1989–1993," *Space Policy* 30, no. 1 (February 1, 2014): 5–12; Michael J. Neufeld, "The Discovery Program: Competition, Innovation, and Risk in Planetary Exploration," in *NASA Spaceflight: A History of Innovation*, ed. Roger D. Launius and Howard E. McCurdy (Cham, Switzerland: Springer International Publishing, 2018), 267–90.

20 Conway, *Exploration and Engineering*, 60.

21 Conway, *Exploration and Engineering*, 101.

22 Conway, *Exploration and Engineering*, 129.

23 Steven J. Dick and James E. Strick, *The Living Universe: NASA and the Development of Astrobiology* (New Brunswick, NJ: Rutgers University Press, 2004), 183.

24 Dick and Strick, *Living Universe*, 106.

25 Conway, *Exploration and Engineering*, 219.

26 W. M. Calvin et al., "Hematite Spherules at Meridiani: Results from MI, Mini-TES, and Pancam," *Journal of Geophysical Research: Planets* 113, no. E12 (2008).

27 Richard V. Morris et al., "Identification of Carbonate-Rich Outcrops on Mars by the Spirit Rover," *Science* 329, no. 5990 (July 23, 2010): 421–24.

28 P. H. Smith et al., "H2O at the Phoenix Landing Site," *Science* 325 (July 1, 2009): 58.

29 E. Heydari et al., "Deposits from Giant Floods in Gale Crater and Their Implications for the Climate of Early Mars," *Scientific Reports* 10, no. 1 (November 5, 2020): 19099.

30 Victor R. Baker, foreword to *Lakes on Mars*, ed. Nathalie A. Cabrol and Edmond A. Grin (Amsterdam: Elsevier, 2010), xviii.

31  Koppes, *JPL and the American Space Program*, 216.

32  Westwick, *Into the Black*, 244.

33  Donna Shirley, *Managing Martians* (New York: Broadway Books, 1998).

34  Conway, *Exploration and Engineering*, 129.

35  Andrew Mishkin, *Sojourner: An Insider's View of the Mars Pathfinder Mission* (New York: Berkley, 2003), 7.

36  Steve Squyres, *Roving Mars: Spirit, Opportunity, and the Exploration of the Red Planet*, reprint edition (New York and London: Hachette Books, 2006), 1.

37  Rob Manning and William L. Simon, *Mars Rover Curiosity: An Inside Account from Curiosity's Chief Engineer* (Washington: Smithsonian Books, 2014).

38  Johnson, *Sirens of Mars*, 179.

39  Andy Weir, *The Martian* (New York: Random House, 2021), 1.

40  The multiplanet world of the Expanse is established in the first book of the series, James S. A. Corey, *Leviathan Wakes* (New York: Orbit, 2011).

41  Silvia Moreno-Garcia, *Prime Meridian* (Vancouver: Innsmouth Free Press, 2017), 82.

### CONCLUSION

1  Jamie Carter, "Moving to Mars: Buzz Aldrin's Vision of Martian Exploration," *BBC Sky at Night Magazine*, September 2017; https://www.skyatnightmagazine.com/space-missions/mars-buzz-aldrin-martian-exploration/.

2  Anthony Pascale, "Interview: Ron Moore on How 'For All Mankind' Is Building the Road to Star Trek," TrekMovie.com, accessed January 26, 2022; https://trekmovie.com/2021/02/15/interview-ron-moore-on-how-for-all-mankind-is-building-the-road-to-star-trek/.

3  "For All Mankind: Ron Moore on His Alt History Space Race Apple+ Series," *Collider*, July 15, 2019; https://collider.com/for-all-mankind-apple-series-explained-ronald-d-moore-interview/.

4  Jeff Foust, "SLS Crawls towards Its First Launch," *Space Review*, March 21, 2022; https://www.thespacereview.com/article/4353/1.

5  James S. J. Schwartz, *The Value of Science in Space Exploration* (Oxford, UK: Oxford University Press, 2020), 27.

6  Michio Kaku, *The Future of Humanity: Terraforming Mars, Interstellar Travel, Immortality, and Our Destiny Beyond Earth* (New York: Random House, 2018), 36.

7  Kaku, *Future of Humanity*, 38.

8  Mike Brown, "SpaceX Mars City: Launch Schedule, Key Build Dates, and How to Get There," *Inverse*, January 6, 2022; https://www.inverse.com/innovation/spacex-mars-city-codex.

9  Robert Zubrin, *Entering Space: Creating a Spacefaring Civilization* (New York: Tarcher/Putnam, 1999), 275.

10  Adam Mann, "Is Mars Ours?" *New Yorker*, accessed May 11, 2021; https://www.newyorker.com/science/elements/is-mars-ours.

11  Christopher Schaberg, "We're Already Colonizing Mars," *Slate*, March 30, 2021; https://slate.com/technology/2021/03/mars-colonization-is-already-happening.html.

12  Fred Scharmen, *Space Forces: A Critical History of Life in Outer Space* (Brooklyn, NY: Verso, 2021).

13  Mann, "Is Mars Ours?"

# *BIBLIOGRAPHY*

Abu-Lughod, Janet L. *Before European Hegemony: The World System A.D. 1250–1350*. New York: Oxford University Press, 1991.

Al-Balkhi, Abu Ma'shar. *The Abbreviation of the Introduction to Astrology: Together with the Medieval Latin Translation of Adelard of Bath*. Translated by Charles Burnett, Keiji Yamamoto, and Michio Yano. New York: E. J. Brill, 1994.

Aldana, Gerardo. *The Apotheosis of Janaab' Pakal: Science, History, and Religion at Classic Maya Palenque*. Boulder: University Press of Colorado, 2010.

Alighieri, Dante. *Il Convito*. Translated by Elizabeth Price Sayer. New York: G. Routledge and Sons, 1887.

An, Liu. *The Essential Huainanzi*. Translated by John S. Major, Sarah A. Queen, Andrew Seth Meyer, and Harold D. Roth. New York: Columbia University Press, 2012.

Asín y Palacios, Miguel. *Islam and the Divine Comedy*. Translated by Harold Sunderland. London: John Murray, 1926.

Aveni, Anthony F. *Ancient Astronomers*. Washington: Smithsonian, 1995.

———. *Skywatchers: A Revised and Updated Version of Skywatchers of Ancient Mexico*. Austin: University of Texas Press, 2001.

———. *Stairways to the Stars: Skywatching in Three Great Ancient Cultures*. New York: Wiley, 1997.

Aveni, Anthony F., Harvey M. Bricker, and Victoria R. Bricker. "Seeking the Sidereal: Observable Planetary Stations and the Ancient Maya Record." *Journal for the History of Astronomy* 34 (May 1, 2003): 145–61.

Baker, Victor R. Foreword to *Lakes on Mars*, edited by Nathalie A. Cabrol and Edmond A. Grin, xv–xx. Amsterdam: Elsevier, 2010.

Barrera-Osorio, Antonio. *Experiencing Nature: The Spanish American Empire and the Early Scientific Revolution*. Austin: University of Texas Press, 2010.

Bazzett, Michael, trans. *The Popol Vuh.* Minneapolis: Milkweed Editions, 2018.

Bimm, Jordan. "Green Mars: Mars Jars and The Origin of Strughold's Vegetation Hypothesis." *Hedgehog,* Fall 2020: 4-7.

Bobrick, Benson. *The Fated Sky: Astrology in History.* New York: Simon & Schuster, 2006.

Boethius, Anicius Manlius Severinus. *The Consolation of Philosophy.* Translated by Richard H. Green. Mineola, NY: Dover Publications, 2002.

Bowler, Peter J. *Evolution: The History of an Idea.* Berkeley: University of California Press, 2009.

Brevig, Eric. Email to Matthew Shindell. "Total Recall Miniature Sets," July 7, 2021.

———. Interview by Matthew Shindell, June 18, 2021.

Bricker, Harvey M., Anthony F. Aveni, and Victoria R. Bricker. "Ancient Maya Documents Concerning the Movements of Mars." *Proceedings of the National Academy of Science* 98 (February 1, 2001): 2107–10.

Bricker, Harvey M., and Victoria R. Bricker. "More on the Mars Table in the Dresden Codex." *Latin American Antiquity* 8, no. 4 (1997): 384–97.

Bricker, Victoria R., and Harvey M. Bricker. "The Seasonal Table in the Dresden Codex and Related Almanacs." *Journal for the History of Astronomy Supplement* 19 (1988): S1.

Briot, D., J. Schneider, and L. Arnold. "G. A. Tikhov, and the Beginnings of Astrobiology." *Extrasolar Planets: Today and Tomorrow* 321 (December 1, 2004): 219–20.

Brooke, John Hedley. *Science and Religion: Some Historical Perspectives.* New York: Cambridge University Press, 1991.

Brown, Mike. "SpaceX Mars City: Launch Schedule, Key Build Dates, and How to Get There." *Inverse,* January 6, 2022. https://www.inverse.com/innovation/spacex-mars-city-codex.

Brush, Stephen G. "The Nebular Hypothesis and the Evolutionary Worldview." *History of Science* 25, no. 3 (September 1, 1987): 245–78.

Byrne, Joseph Patrick. *Daily Life during the Black Death.* Westport, CT: Greenwood Publishing Group, 2006.

Calvin, W. M., J. D. Shoffner, J. R. Johnson, A. H. Knoll, J. M. Pocock, S. W. Squyres, C. M. Weitz, et al. "Hematite Spherules at Meridiani: Results from MI, Mini-TES, and Pancam." *Journal of Geophysical Research: Planets* 113, no. E12 (2008).

Camenietzki, Carlos Ziller. "Baroque Science between the Old and the New World: Father Kircher and His Colleague Valentin Stansel (1621–1705)." In *Athanasius Kircher: The Last Man Who Knew Everything,* edited by Paula Findlen, 311–28. New York: Routledge, 2004.

Campbell, Mary Baine. *Wonder and Science: Imagining Worlds in Early Modern Europe.* Ithaca, NY: Cornell University Press, 2004.

Campion, Nicholas. *A History of Western Astrology, Volume 1: The Ancient World.* New York: Bloomsbury Academic, 2008.

Capella, Martianus. *Martianus Capella and the Seven Liberal Arts: The Marriage of Philology and Mercury.* Translated by William Harris Stahl. New York: Columbia University Press, 1971.

Carr, Michael H. *The Surface of Mars.* New Haven, CT: Yale University Press, 1981.

Carter, Jamie. "Moving to Mars: Buzz Aldrin's Vision of Martian Exploration." *BBC Sky at Night Magazine,* September 2017. https://www.skyatnightmagazine.com/space-missions/mars-buzz-aldrin-martian-exploration/.

Chambers, Robert. *Vestiges of the Natural History of Creation and Other Evolutionary Writings.* Edited by James A. Secord. Chicago: University of Chicago Press, 1994.

Christensen, P. R., J. L. Bandfield, R. N. Clark, K. S. Edgett, V. E. Hamilton, T. Hoefen, H. H. Kieffer, et al. "Detection of Crystalline Hematite Mineralization on Mars by the Thermal Emission Spectrometer: Evidence for Near-Surface Water." *Journal of Geophysical Research: Planets* 105, no. E4 (2000): 9623–42.

Christensen, Philip R., and Matthew B. Shindell. "Mars Infrared." *Planetary Report*, June 2003.

Clancey, William J. *Working on Mars: Voyages of Scientific Discovery with the Mars Exploration Rovers*. Cambridge, MA: MIT Press, 2012.

Coe, Michael D., and Mark Van Stone. *Reading the Maya Glyphs*. London: Thames & Hudson, 2005.

Conway, Erik M. *Exploration and Engineering: The Jet Propulsion Laboratory and the Quest for Mars*. Baltimore: Johns Hopkins University Press, 2015.

Corey, James S. A. *Leviathan Wakes*. New York: Orbit, 2011.

Crossley, Robert. *Imagining Mars: A Literary History*. Middletown, CT: Wesleyan University Press, 2011.

Cullen, Christopher. *Heavenly Numbers: Astronomy and Authority in Early Imperial China*. New York: Oxford University Press, 2017.

Dear, Peter. *Revolutionizing the Sciences: European Knowledge in Transition, 1500–1700*. 3rd ed. London: Red Globe Press, 2019.

Dekker, Elly. "Globes in Renaissance Europe." In *The History of Cartography, Volume 3*, edited by David Woodward, 135–73. Chicago: University of Chicago Press, 2007.

DeVorkin, David. *Science with a Vengeance: How the Military Created the US Space Sciences after World War II*. New York: Springer-Verlag, 1992.

Dick, Philip K. "We Can Remember It for You Wholesale." In *The Wesleyan Anthology of Science Fiction*, edited by Arthur B. Evans, Istvan Csicsery-Ronay, Joan Gordon, Veronica Hollinger, Rob Latham, and Carol McGuirk, 385–404. Middletown, CT: Wesleyan University Press, 2010.

Dick, Steven J. *The Biological Universe: The Twentieth Century Extraterrestrial Life Debate and the Limits of Science*. New York: Cambridge University Press, 1996.

Dick, Steven J., and James E. Strick. *The Living Universe: NASA and the Development of Astrobiology*. New Brunswick, NJ: Rutgers University Press, 2004.

Eisfeld, Rainer. "Projecting Landscapes of the Human Mind onto Another World: Changing Faces of an Imaginary Mars." In *Radical Approaches to Political Science: Roads Less Traveled*, 185–204. Berlin: Barbara Budrich, 2012.

Ezell, Edward Clinton, and Linda Neuman Ezell. *On Mars: Exploration of the Red Planet, 1958–1978*. Washington: National Aeronautics and Space Administration, 1984.

Filiatrault, Jason. "Goodbye, Opportunity Rover." *Washington Post*, February 13, 2019. https://www.washingtonpost.com/opinions/2019/02/13/goodbye-opportunity-rover-thank-you-letting-humanity-see-mars-with-your-eyes/.

Finocchiaro, Maurice A., ed. *The Galileo Affair: A Documentary History*. Berkeley: University of California Press, 1989.

Fletcher, John Edward. *A Study of the Life and Works of Athanasius Kircher, 'Germanus Incredibilis.'* Boston: Brill, 2011.

Foust, Jeff. "Red Planet Scare." *Space Review*, May 24, 2021. https://thespacereview.com/article/4181/1.

———. "SLS Crawls towards Its First Launch." *Space Review*, March 21, 2022. https://www.thespacereview.com/article/4353/1.

Freidel, David, Linda Schele, and Joy Parker. *Maya Cosmos: Three Thousand Years on the Shaman's Path*. New York: HarperCollins Publishers, 2001.

Geller, Markham Judah. *Melothesia in Babylonia: Medicine, Magic, and Astrology in the Ancient Near East*. Boston: Walter de Gruyter, 2014.

Glomski, Jacqueline. "Religion, the Cosmos, and Counter-Reformation Latin: Athanasius Kircher's Itinerarium Exstaticum (1656)." *Acta Conventus Neo-Latini Monasteriensis: Proceedings of the Fifteenth International Congress of Neo-Latin Studies (Münster 2012)*, March 2015, 227–36.

Goldstein, Bernard R. "Astronomy and the Jewish Community in Early Islam." *Aleph*, no. 1 (2001): 17–57.

———. "Astronomy as a 'Neutral Zone': Interreligious Cooperation In Medieval Spain." *Medieval Encounters* 15, no. 2 (2009): 159–74.

———. "The Making of Astronomy in Early Islam." *Nuncius* 1, no. 2 (January 1, 1986): 79–92.

Gómez, Nicolás Wey. *The Tropics of Empire: Why Columbus Sailed South to the Indies*. Cambridge, MA: MIT Press, 2008.

Grafton, Anthony, April Shelford, and Nancy Siraisi. *New Worlds, Ancient Texts: The Power of Tradition and the Shock of Discovery*. Cambridge, MA: Harvard University Press, 1995.

Grant, Edward. "Cosmology." In *The Cambridge History of Science, Vol. 2: Medieval Science*, edited by David C. Lindberg and Michael H. Shank, 436-55. Cambridge, UK: Cambridge University Press, 2013.

———. *The Foundations of Modern Science in the Middle Ages: Their Religious, Institutional and Intellectual Contexts*. New York: Cambridge University Press, 1996.

———. "The Medieval Cosmos: Its Structure and Operation." *Journal for the History of Astronomy* 28, no. 2 (May 1, 1997): 147–67.

———. *Planets, Stars, and Orbs: The Medieval Cosmos, 1200–1687*. New York: Cambridge University Press, 1994.

Grant, John. "Gaining Inspiration from *The Martian Chronicles*." *Stories from the Smithsonian National Air and Space Museum* (blog), August 20, 2020. https://airandspace.si.edu/stories/editorial/gaining-inspiration-martian-chronicles.

Greg, Percy. *Across the Zodiac: The Story of a Wrecked Record*. London: Trübner, 1880.

Greicius, Tony. "The Launch Is Approaching for NASA's Next Mars Rover, Perseverance." NASA, June 17, 2020. http://www.nasa.gov/feature/jpl/the-launch-is-approaching-for-nasas-next-mars-rover-perseverance.

Gutas, Dimitri. *Greek Thought, Arabic Culture: The Graeco-Arabic Translation Movement in Baghdad and Early 'Abbasid Society (2nd–4th / 5th–10th c.)*. London: Routledge, 2012.

Guthke, Karl S. *Exploring the Interior: Essays on Literary and Cultural History*. Cambridge, UK: Open Book Publishers, 2018.

———. *The Last Frontier: Imagining Other Worlds from the Copernican Revolution to Modern Science Fiction*. Ithaca, NY: Cornell University Press, 1990.

Hahn, Roger. "Laplace and the Mechanical Universe." In *God and Nature: Historical Essays on the Encounter between Christianity and Science*, edited by David C. Lindberg and Ronald L. Numbers, 256–95. Berkeley: University of California Press, 1986.

Heilbron, John L. *Galileo*. Oxford, UK: Oxford University Press, 2012.

Henry, John. "Primary and Secondary Causation in Samuel Clarke's and Isaac Newton's Theories of Gravity." *Isis* 111, no. 3 (September 1, 2020): 542–61.

Heydari, E., J. F. Schroeder, F. J. Calef, J. Van Beek, S. K. Rowland, T. J. Parker, and A. G. Fairén. "Deposits from Giant Floods in Gale Crater and Their Implications for the Climate of Early Mars." *Scientific Reports* 10, no. 1 (November 5, 2020): 190–99.

Hobson, John M. *The Eastern Origins of Western Civilisation*. New York: Cambridge University Press, 2004.

Hockey, Thomas. *How We See the Sky: A Naked-Eye Tour of Day and Night*. Chicago: University of Chicago Press, 2011.

Horodowich, Elizabeth. "Italy and the New World." In *The New World in Early Modern Italy, 1492–1750*, edited by Elizabeth Horodowich and Lia Markey, 19–33. New York: Cambridge University Press, 2018.

———. *The Venetian Discovery of America: Geographic Imagination and Print Culture in the Age of Encounters*. New York: Cambridge University Press, 2018.

Horrox, Rosemary. *The Black Death*. Manchester, UK: Manchester University Press, 1994.

Hoyt, William Graves. *Lowell and Mars*. Tucson: University of Arizona Press, 1996.

Huntress Jr., Wesley T., and Mikhail Ya Marov. *Soviet Robots in the Solar System: Mission Technologies and Discoveries*. New York: Springer, 2011.

Hutchinson, Henry Neville. *The Autobiography of the Earth: A Popular Account of Geological History*. London: Edward Stanford, 1890.

Isserles, Justine. "Bloodletting and Medical Astrology in Hebrew Manuscripts from Medieval Western Europe." *Sudhoffs Archiv* 101, no. 1 (2017): 2–41.

Jet Propulsion Laboratory (US). *Mariner Mission to Venus: Prepared for the National Aeronautics and Space Administration*. New York: McGraw-Hill, 1963.

Johnson, Sarah Stewart. *The Sirens of Mars: Searching for Life on Another World*. New York: Crown, 2020.

Kaku, Michio. *The Future of Humanity: Terraforming Mars, Interstellar Travel, Immortality, and Our Destiny Beyond Earth*. New York: Random House, 2018.

Kauntze, Mark. *Authority and Imitation: A Study of the* Cosmographia *of Bernard Silvestris*. Boston: Brill, 2014.

Kemp, Martin. "Moving in Elevated Circles." *Nature* 466, no. 7302 (July 2010): 33.

Kieckhefer, Richard. *Forbidden Rites: A Necromancer's Manual of the Fifteenth Century*. Stroud, UK: Sutton, 1997.

Kilgore, De Witt Douglas. *Astrofuturism: Science, Race, and Visions of Utopia in Space*. Philadelphia: University of Pennsylvania Press, 2003.

Kiviat, Robert C. "Casting a New Light on the Mars Face." *Omni* 16, no. 11 (August 1, 1994): 31–37.

Koppes, Clayton R. *JPL and the American Space Program: A History of the Jet Propulsion Laboratory*. New Haven, CT: Yale University Press, 1982.

Kraemer, Joel L. *Humanism in the Renaissance of Islam: The Cultural Revival during the Buyid Age*. Leiden, Netherlands: E. J. Brill, 1986.

Lambert, W. G. "*Enuma Elish*: The Babylonian Epic of Creation." *World History Encyclopedia*. Accessed April 10, 2021. https://www.ancient.eu/article/225/enuma-elish—-the-babylonian -epic-of-creation—-fu/.

Lambright, W. Henry. "Big Science in Space: Viking, Cassini, and the Hubble Space Telescope." In *Exploring the Solar System: The History and Science of Planetary Exploration*, edited by Roger D. Launius, 129–48. New York: Palgrave Macmillan, 2013.

———. *Why Mars: NASA and the Politics of Space Exploration*. Baltimore: Johns Hopkins University Press, 2014.

Lane, K. Maria D. "Geographers of Mars: Cartographic Inscription and Exploration Narrative in Late Victorian Representations of the Red Planet." *Isis* 96, no. 4 (2005): 477–506.

———. "Mapping the Mars Canal Mania: Cartographic Projection and the Creation of a Popular Icon." *Imago Mundi* 58, no. 2 (2006): 198–211.

Lichtman, Flora, and Sharon Shattuck. "Animated Life: Pangea." *New York Times*, February 17, 2015. https://www.nytimes.com/video/opinion/100000003515124/animated-life-pangaea.html.

Lidaka, Juris. "The Book of Angels, Rings, Characters, and Images of the Planets: Attributed to Osbern Bokenham." In *Conjuring Spirits: Texts and Traditions of Late Medieval Ritual Magic*, edited by Claire Fanger, 32–75. University Park: Pennsylvania State University Press, 1994.

Lightman, Bernard. *Victorian Popularizers of Science: Designing Nature for New Audiences*. Chicago: University of Chicago Press, 2007.

———. "The Visual Theology of Victorian Popularizers of Science: From Reverent Eye to Chemical Retina." *Isis* 91, no. 4 (2000): 651–80.

———. "'The Voices of Nature': Popularizing Victorian Science." In *Victorian Science in Context*, 187–211. Chicago: University of Chicago Press, 1997.

Lindberg, David C. *The Beginnings of Western Science: The European Scientific Tradition in Philosophical, Religious, and Institutional Context, Prehistory to A.D. 1450*. Chicago: University of Chicago Press, 2008.

Logsdon, John M. *After Apollo? Richard Nixon and the American Space Program*. New York: Palgrave Macmillan, 2015.

Macrobius, Ambrosius Aurelius Theodosius. *Commentary on the Dream of Scipio*. Translated by William Harris Stahl. New York: Columbia University Press, 1990.

Major, John S. *Heaven and Earth in Early Han Thought: Chapters Three, Four, and Five of the Huainanzi*. Albany: State University of New York Press, 1993.

Mann, Adam. "Is Mars Ours?" *New Yorker*. Accessed May 11, 2021. https://www.newyorker.com/science/elements/is-mars-ours.

Manning, Rob, and William L. Simon. *Mars Rover Curiosity: An Inside Account from Curiosity's Chief Engineer*. Washington: Smithsonian Books, 2014.

Margolis, Jacob. "How a Tweet about the Mars Rover Dying Blew Up on the Internet and Made People Cry." *LAist*, February 16, 2019. https://laist.com/news/jpl-mars-rover-opportunity-battery-is-low-and-its-getting-dark.

Markey, Lia. *Imagining the Americas in Medici Florence*. University Park: Pennsylvania State University Press, 2016.

Markley, Robert. *Dying Planet: Mars in Science and the Imagination*. Durham, NC: Duke University Press, 2005.

———. "Missions to Mars: Reimagining the Red Planet in the Age of Spaceflight." In *Exploring the Solar System: The History and Science of Planetary Exploration*, edited by Roger D. Launius, 249–72. New York: Palgrave Macmillan, 2013.

*Mars as Viewed by Mariner 9: A Pictorial Presentation by the Mariner 9 Television Team and the Planetology Program Principal Investigators*. Washington: National Aeronautics and Space Administration, 1974.

McCray, W. Patrick. *The Visioneers: How a Group of Elite Scientists Pursued Space Colonies, Nanotechnologies, and a Limitless Future*. Princeton, NJ: Princeton University Press, 2013.

Messeri, Lisa. *Placing Outer Space: An Earthly Ethnography of Other Worlds*. Durham, NC: Duke University Press, 2016.

Miller, Genevieve. "'Airs, Waters, and Places' in History." *Journal of the History of Medicine and Allied Sciences* 17, no. 1 (January 1, 1962): 129–40.

Mirmalek, Zara. *Making Time on Mars.* Cambridge, MA: MIT Press, 2020.

Mishkin, Andrew. *Sojourner: An Insider's View of the Mars Pathfinder Mission.* New York: Berkley Publishing Group, 2003.

Mithen, Steven. *The Prehistory of the Mind: A Search for the Origins of Art, Religion and Science.* London: Orion Books, 1998.

Moreno-Garcia, Silvia. *Prime Meridian.* Vancouver: Innsmouth Free Press, 2017.

Morris, Richard V., Steven W. Ruff, Ralf Gellert, Douglas W. Ming, Raymond E. Arvidson, Benton C. Clark, D. C. Golden, et al. "Identification of Carbonate-Rich Outcrops on Mars by the Spirit Rover." *Science* 329, no. 5990 (July 23, 2010): 421–24.

Morrison, Robert G. "Islamic Astronomy." In *The Cambridge History of Science, Volume 2: Medieval Science,* edited by David C. Lindberg and Michael H. Shank, 109–38. Cambridge, UK: Cambridge University Press, 2013.

Morrissey, Thomas J. "Ready or Not, Here We Come: Metaphors of the Martian Megatext from Wells to Robinson." *Journal of the Fantastic in the Arts* 10, no. 4 (40) (2000): 372–94.

Moskowitz, Sam, ed. *Under the Moons of Mars: A History and Anthology of The Scientific Romance in the Munsey Magazines, 1912–1920.* New York: Holt Reinhart & Winston, 1970.

Mukerji, Chandra. *A Fragile Power: Scientists and the State.* Princeton, NJ: Princeton University Press, 1989.

Murray, Bruce, Michael C. Malin, and Ronald Greeley. *Earthlike Planets: Surfaces of Mercury, Venus, Earth, Moon, Mars.* San Francisco: W. H. Freeman, 1981.

Mutch, Thomas A., Raymond E. Arvidson, James W. Head, Kenneth L. Jones, and R. Stephen Saunders. *The Geology of Mars.* Princeton, NJ: Princeton University Press, 1976.

Nall, Joshua. *News from Mars: Mass Media and the Forging of a New Astronomy, 1860–1910.* Pittsburgh: University of Pittsburgh Press, 2019.

Neufeld, Michael. *Von Braun: Dreamer of Space, Engineer of War.* New York: Vintage Books, 2008.

Neufeld, Michael J. "The Discovery Program: Competition, Innovation, and Risk in Planetary Exploration." In *NASA Spaceflight: A History of Innovation,* edited by Roger D. Launius and Howard E. McCurdy, 267–90. Cham, Switzerland: Springer International Publishing, 2018.

———. "Transforming Solar System Exploration: The Origins of the Discovery Program, 1989–1993." *Space Policy* 30, no. 1 (February 1, 2014): 5–12.

Numbers, Ronald L. *Creation by Natural Law: Laplace's Nebular Hypothesis in American Thought.* Seattle: University of Washington Press, 1977.

Nunn, George E. "The Imago Mundi and Columbus." *American Historical Review* 40, no. 4 (July 1, 1935): 646–61.

Ogilvie, Marilyn Bailey. "Robert Chambers and the Nebular Hypothesis." *British Journal for the History of Science* 8, no. 3 (1975): 214–32.

Oreskes, Naomi. *Science on a Mission: How Military Funding Shaped What We Do and Don't Know about the Ocean.* Chicago: University of Chicago Press, 2021.

Pascale, Anthony. "Interview: Ron Moore on How For All Mankind Is Building the Road to Star Trek." *TrekMovie.com.* Accessed January 26, 2022. https://trekmovie.com/2021/02/15/interview-ron-moore-on-how-for-all-mankind-is-building-the-road-to-star-trek/.

Pieri, David. Interview by Matthew Shindell, July 10, 2021.

Poskett, James. *Horizons: The Global Origins of Modern Science*. Boston: Mariner Books, 2022.

Principe, Lawrence M. *The Secrets of Alchemy*. Illustrated edition. Chicago: University of Chicago Press, 2015.

Ptolemaeus, Claudius. *Tetrabiblos*. Translated by Frank E. Robbins. Cambridge, MA: Harvard University Press, 1940.

Radish, Christina. "*For All Mankind*: Ron Moore on His Alt History Space Race Apple+ Series," *Collider*, July 15, 2019. https://collider.com/for-all-mankind-apple-series-explained-ronald-d-moore-interview/.

Rice, Eugene F., and Anthony Grafton. *The Foundations of Early Modern Europe, 1460–1559*. 2nd ed. New York: W. W. Norton, 1994.

Robbins, Lawrence H. "Astronomy and Prehistory." In *Astronomy across Cultures: The History of Non-Western Astronomy*, edited by Helaine Selin, 31–52. Boston: Kluwer Academic Publishers, 2000.

Roberts, Paul. "Ego Trip." *Cinefex* 43 (August 1990): 4–33.

Robinson, Kim Stanley. *Red Mars*. New York: Random House Publishing Group, 2003.

Rochberg, Francesca. *The Heavenly Writing: Divination, Horoscopy, and Astronomy in Mesopotamian Culture*. New York: Cambridge University Press, 2004.

———. *In the Path of the Moon: Babylonian Celestial Divination and Its Legacy*. Boston: Brill, 2010.

Rosen, Mark. "A New Chronology of the Construction and Restoration of the Medici Guardaroba in the Palazzo Vecchio, Florence." *Mitteilungen Des Kunsthistorischen Institutes in Florenz* 53, no. 2/3 (2009): 285–308.

Rowland, Ingrid. "'Th' United Sense of th' Universe': Athanasius Kircher in Piazza Navona." *Memoirs of the American Academy in Rome* 46 (2001): 153–81.

Rowland, Ingrid D. "Athanasius Kircher, Giordano Bruno, and the Panspermia of the Infinite Universe." In *Athanasius Kircher: The Last Man Who Knew Everything*, edited by Paula Findlen, 191–205. New York: Routledge, 2004.

———. *The Ecstatic Journey: Athanasius Kircher in Baroque Rome*. Chicago: University of Chicago Library, 2000.

———. "Poetry and Prophecy in the Encyclopedic System of Athanasius Kircher." *Bruniana & Campanelliana* 11, no. 2 (2005): 509–17.

Ryan, Michael A. *A Kingdom of Stargazers: Astrology and Authority in the Late Medieval Crown of Aragon*. Ithaca, NY: Cornell University Press, 2011.

Sabra, A. I. "The Appropriation and Subsequent Naturalization of Greek Science in Medieval Islam: A Preliminary Statement." *History of Science* 25, no. 3 (September 1, 1987): 223–43.

———. "Situating Arabic Science: Locality versus Essence." *Isis* 87, no. 4 (1996): 654–70.

Sagan, Carl. *Cosmos*. New York: Ballantine Books, 2013.

Sawyer, Kathy. "A Mars Never Dreamed Of." *National Geographic*, February 2001. https://www.nationalgeographic.com/science/article/mars-explored-anew.

Schaberg, Christopher. "We're Already Colonizing Mars." *Slate*, March 30, 2021. https://slate.com/technology/2021/03/mars-colonization-is-already-happening.html.

Schaffer, Simon. "'The Great Laboratories of the Universe': William Herschel on Matter Theory and Planetary Life." *Journal for the History of Astronomy* 11 (1980): 81–110.

———. "Herschel in Bedlam: Natural History and Stellar Astronomy." *British Journal for the History of Science* 13, no. 3 (1980): 211–39.

Scharmen, Fred. *Space Forces: A Critical History of Life in Outer Space*. Brooklyn, NY: Verso, 2021.

Schele, Linda, and Mary Ellen Miller. *Blood of Kings: Dynasty and Ritual in Maya Art.* New York: George Braziller, 1986.

Schwartz, James S. J. *The Value of Science in Space Exploration.* Oxford, UK: Oxford University Press, 2020.

Secord, James A. *Victorian Sensation: The Extraordinary Publication, Reception, and Secret Authorship of Vestiges of the Natural History of Creation.* Chicago: University of Chicago Press, 2000.

Sheehan, William, and Jim Bell. *Discovering Mars: A History of Observation and Exploration of the Red Planet.* Tucson: University of Arizona Press, 2021.

Shindell, Matthew. "Domesticating the Planets: Instruments and Practices in the Development of Planetary Geology." *Spontaneous Generations: A Journal for the History and Philosophy of Science* 4, no. 1 (2010): 191–230.

———. "From the End of the World to the Age of the Earth: The Cold War Development of Isotope Geochemistry at the University of Chicago and Caltech." In *Science and Technology in the Global Cold War*, edited by Naomi Oreskes and John Krige. Cambridge, MA: MIT Press, 2014.

———. "Geophysics." In *A Companion to the History of American Science*, 120–33. Boston: John Wiley & Sons, 2015.

Shirley, Donna. *Managing Martians.* New York: Broadway Books, 1998.

Siebert, Harald. "The Early Search for Stellar Parallax: Galileo, Castelli, and Ramponi." *Journal for the History of Astronomy* 36 (August 1, 2005): 251–71.

Silvestris, Bernardus. *The Cosmographia of Bernardus Silvestris.* Translated by Winthrop Wetherbee. New York: Columbia University Press, 1990.

Sinton, William M. "Further Evidence of Vegetation on Mars." *Science* 130, no. 3384 (1959): 1234–37.

Sivin, Nathan. *Granting the Seasons: The Chinese Astronomical Reform of 1280, with a Study of Its Many Dimensions and a Translation of Its Records.* New York: Springer, 2009.

Smith, P. H., L. K. Tamppari, R. E. Arvidson, D. Bass, D. Blaney, W. V. Boynton, A. Carswell, et al. "$H_2O$ at the Phoenix Landing Site." *Science* 325 (July 1, 2009): 58.

Spufford, Francis. *I May Be Some Time: Ice and the English Imagination.* New York: St. Martin's Press, 1997.

Squyres, Steve. *Roving Mars: Spirit, Opportunity, and the Exploration of the Red Planet.* Reprint edition. New York and London: Hachette Books, 2006.

Stetson, Mark. Interview by Matthew Shindell, July 11, 2021.

Stolzenberg, Daniel. *Egyptian Oedipus: Athanasius Kircher and the Secrets of Antiquity.* Chicago: University of Chicago Press, 2013.

Swerdlow, Noel M. *The Babylonian Theory of the Planets.* Princeton, NJ: Princeton University Press, 1998.

Tanaka, Kenneth L. "The Stratigraphy of Mars." *Journal of Geophysical Research* 91, no. B13 (November 30, 1986): E139–58.

Taylor, Kenneth L. "Earth and Heaven, 1750–1800: Enlightenment Ideas about the Relevance to Geology of Extraterrestrial Operations and Events." *Earth Sciences History* 17, no. 2 (1998): 84–91.

Tesorero, Angel. "Meet the Emirati Engineers of Hope Probe Mars Mission." *Gulf News*, February 10, 2021. https://gulfnews.com/special-reports/meet-the-emirati-engineers-of -hope-probe-mars-mission-1.77094280.

Tomayko, James. *Computers in Spaceflight: The NASA Experience*. Washington: National Aeronautics and Space Administration, 1988.

Tseng, Lillian Lan-ying. *Picturing Heaven in Early China*. Cambridge, MA: Harvard University Press, 2011.

Vertesi, Janet. *Seeing Like a Rover: How Robots, Teams, and Images Craft Knowledge of Mars*. Chicago: University of Chicago Press, 2015.

Waddell, Mark A. *Magic, Science, and Religion in Early Modern Europe*. New Approaches to the History of Science and Medicine. New York: Cambridge University Press, 2021.

Wall, Mike. "NASA's Next Mars Rover Carries Tribute to Healthcare Workers Fighting Coronavirus." *Space.com*, June 17, 2020. https://www.space.com/nasa-mars-rover-perseverance-coronavirus-tribute.html.

Weir, Andy. *The Martian*. New York: Random House, 2021.

Westfall, Richard S. *The Construction of Modern Science: Mechanisms and Mechanics*. New York: Cambridge University Press, 1977.

———. *The Life of Isaac Newton*. New York: Cambridge University Press, 1993.

Westman, Robert. *The Copernican Question: Prognostication, Skepticism, and Celestial Order*. Berkeley: University of California Press, 2011.

Westwick, Peter J. *Into the Black: JPL and the American Space Program, 1976–2004*. New Haven, CT: Yale University Press, 2007.

Willis, Martin. *Vision, Science and Literature, 1870–1920: Ocular Horizons*. Pittsburgh, PA: University of Pittsburgh Press, 2016.

Wilson, Albert G., Hubertus Strughold, William M. Sinton, Audouin Dollfus, John A. Kooistra, Roland B. Mitchell, David G. Simons, and Ingeborg Schmidt. "Problems Common to the Fields of Astronomy and Biology: A Symposium." *Publications of the Astronomical Society of the Pacific* 70, no. 412 (1958): 41–78.

Wilson, Miles. *The History of Israel Jobson, the Wandering Jew*. London: J. Nicholson, 1757.

Xiaochun, Sun. "Crossing the Boundaries between Heaven and Man: Astronomy in Ancient China." In *Astronomy across Cultures: The History of Non-Western Astronomy*, edited by Helaine Selin and Sun Xiaochun, 423–54. Boston: Springer, 2000.

Zubrin, Robert. *Entering Space: Creating a Spacefaring Civilization*. New York: Tarcher/Putnam, 1999.

# INDEX